CHILTON'S Guide to Telephone Installation and Repair

CHILTON'S Guide to Telephone Installation and Repair

JOHN T. MARTIN

CHILTON BOOK COMPANY RADNOR, PENNSYLVANIA

Published in Radnor, Pennsylvania 19089 by Chilton Book Company

Designed by William E. Lickfield
Manufactured in the United States of America

Library of Congress Cataloging in Publication Data
Martin, John T.
Chilton's guide to telephone installation
and repair.
Includes index.
1. Telephone—Amateurs' manuals. I. Title.
II. Title: Guide to telephone installation and
repair.
TK9951.M325 1985 621.386 84-45692
ISBN 0-8019-7602-2 (pbk.)

2 3 4 5 6 7 8 9 0 4 3 2 1 0 9 8 7 6

Contents

Preface

Ever since the telephone was invented, consumers have had but one source to turn to for complete information about home telephones: the telephone company. Today, the consumer is offered a second source: CHILTON'S GUIDE TO TELEPHONE INSTALLATION AND REPAIR.

The goal of this book is to provide you, the home telephone consumer, with sufficient information to answer any and all questions about your home telephone system—even questions you might not have known to ask. The information included has been gleaned from years of working experience in the home telecommunications field in a number of positions with a variety of telephone companies. Because the book details every step of installation and repair, leaving nothing to chance, it will be beneficial even to the non-do-it-yourself person.

Acknowledgments

I would like to thank my mother

BYRD HARRIS MARTIN

without whose support this book would not have been possible.

I would also like to thank my good friend

BARBARA E. BENNETT

who labored many hours typing my final manuscript.

Telephone wiring may contain varying amounts of electrical current, ranging from 48 volts DC to 105 volts AC. Although electrical shock at these voltages is usually not harmful, the following precautions should be taken:

1. Never work on your telephone system during a thunderstorm.
2. If you use a pacemaker, do not work on your telephone system.
3. Use insulated tools or wear heavy rubber gloves.
4. Don't touch grounded objects or appliances while working on your telephone system.
5. If you have a telephone at a location other than the one where you are working, take its receiver off the hook. This will cause your telephone circuit to be busy so that incoming calls will not produce ringing voltage, which is much higher than normal line voltage.
6. If you have only one telephone, take the receiver off the hook while you are working on it.
7. If you drill through existing walls, be careful not to come in contact with electrical wiring or hidden plumbing pipes.

CHILTON'S Guide to Telephone Installation and Repair

INTRODUCTION

General Description of a Telephone System

Understanding the fundamental concepts, operating characteristics, and terminology relating to your home telephone will enable you to obtain the most from it. Learning how a telephone system works is the best place to begin. A telephone *system* refers to your telephone and all the steps it goes through in placing and receiving a call.

The Pair

A telephone system uses only two wires, known as a *pair,* to place and receive calls. Ninety percent of the working operation of your telephone system deals with the pair and the connecting points where the wires travel to place and receive calls.

Figure I-1 gives a simple description of a telephone system, which is almost no more complicated than a child's tin can and string telephone. (If you find yourself becoming confused later in the book refer back to this illustration.)

To understand the actual working operation of your telephone system, you will need to know how the pair connects your telephone with another.

Telephone Cables

Telephone *cables* are pairs that are bound together; each telephone's pair is connected with other pairs through a cable network.

In order to keep track of each individual pair, the telephone company uses a standard color code for labeling. This method enables the company to assign and locate a pair at any point within the telephone cable network.

Basically, all telephone cables are alike even though their names vary according to their location. For example, cables placed on telephone poles are referred to as aerial telephone cables; cables placed underground are called buried telephone cables; and cables that span between

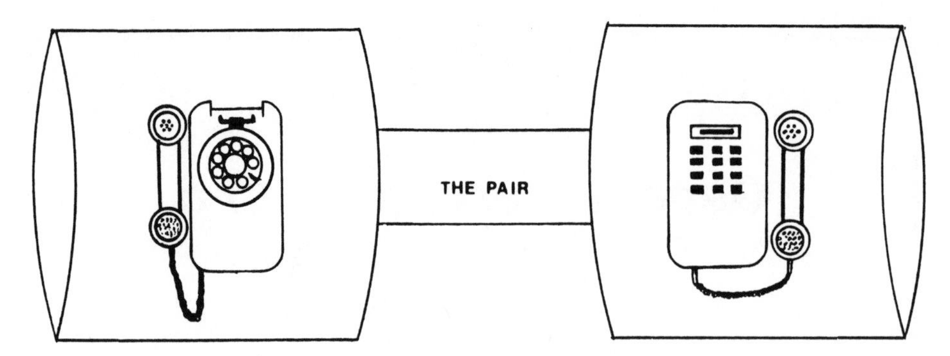

Fig. I-1 A telephone system.

the central office and your home telephone are known as local-network telephone cables. Figure I-2 shows the local telephone network and the specific names given to the telephone cables within it.

The Local Network

Each telephone system is a part of a *local network,* which is owned and operated by your local telephone company. The local network consists of a permanent structure of telephone cables, switching equipment, and terminals that provide the fundamental link between your telephone and all others.

The quality of your telephone service is greatly affected by the quality of the local-network operation. The following pages discuss in detail each connecting point within the local network from the central office to your home telephone. (Refer to Figure I-2 to see the relationship between network components.)

The Central Office

The *central office* (C.O.) is at one end of the local network while your home telephone is at the other end.

As the originating location and nucleus of your telephone system, the central office is the most complex area of the local network, yet all the home telephone user really needs to know about the central office is what functions it provides to the telephone system.

Figure I-3 shows a typical central office. Switching equipment at the central office directs your calls to their desired locations. Other equipment supplies the dial tone, the busy signal, and the electrical current necessary for voice transmission. The ringing generator supplies the ring. Some service options depend on what types of supplemental equipment is installed at the central office.

The Cables

The telephone cable that connects one central office with another in a different local network is called trunk cable. Main cable and feeder cable

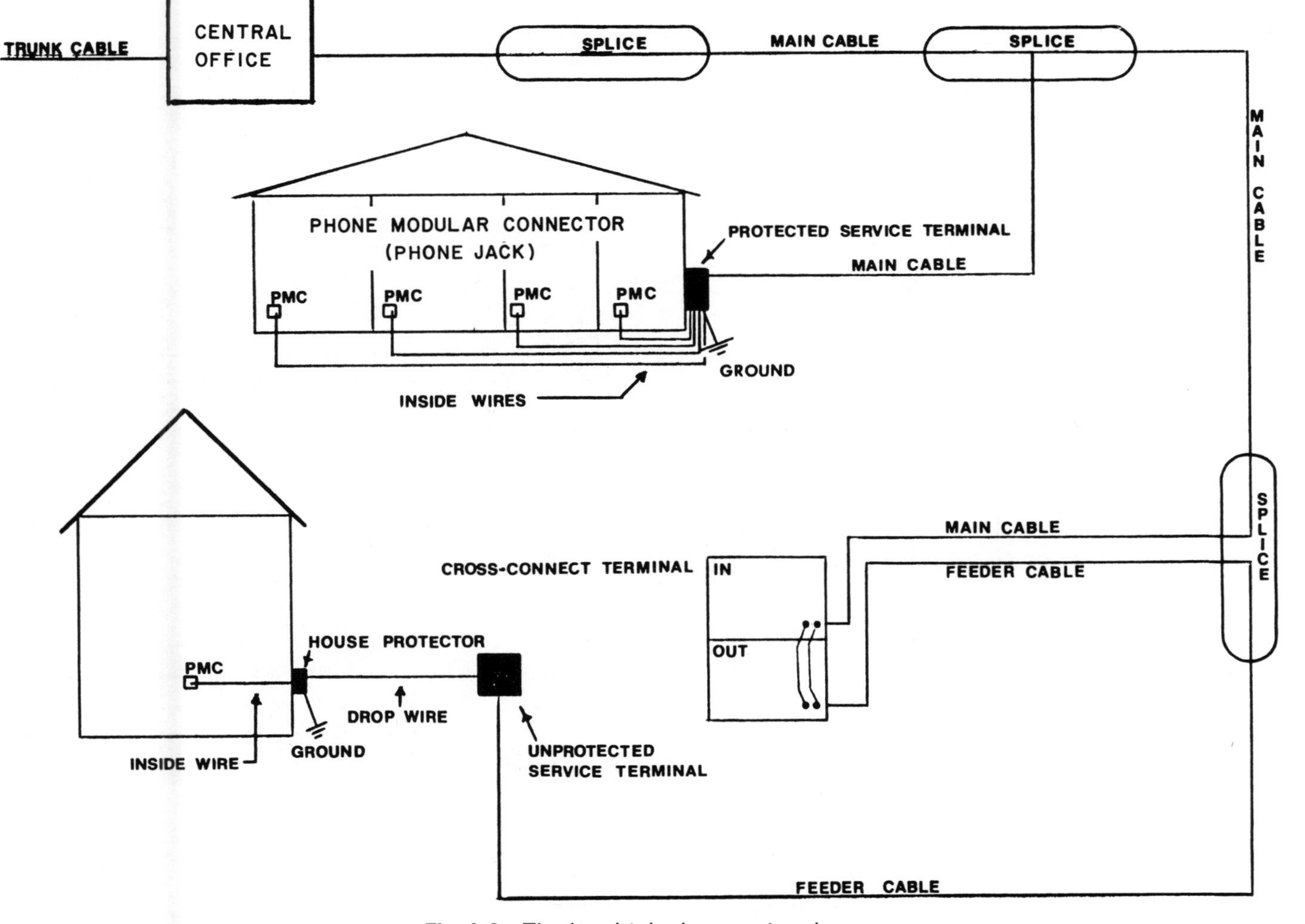

Fig. I-2 The local telephone network.

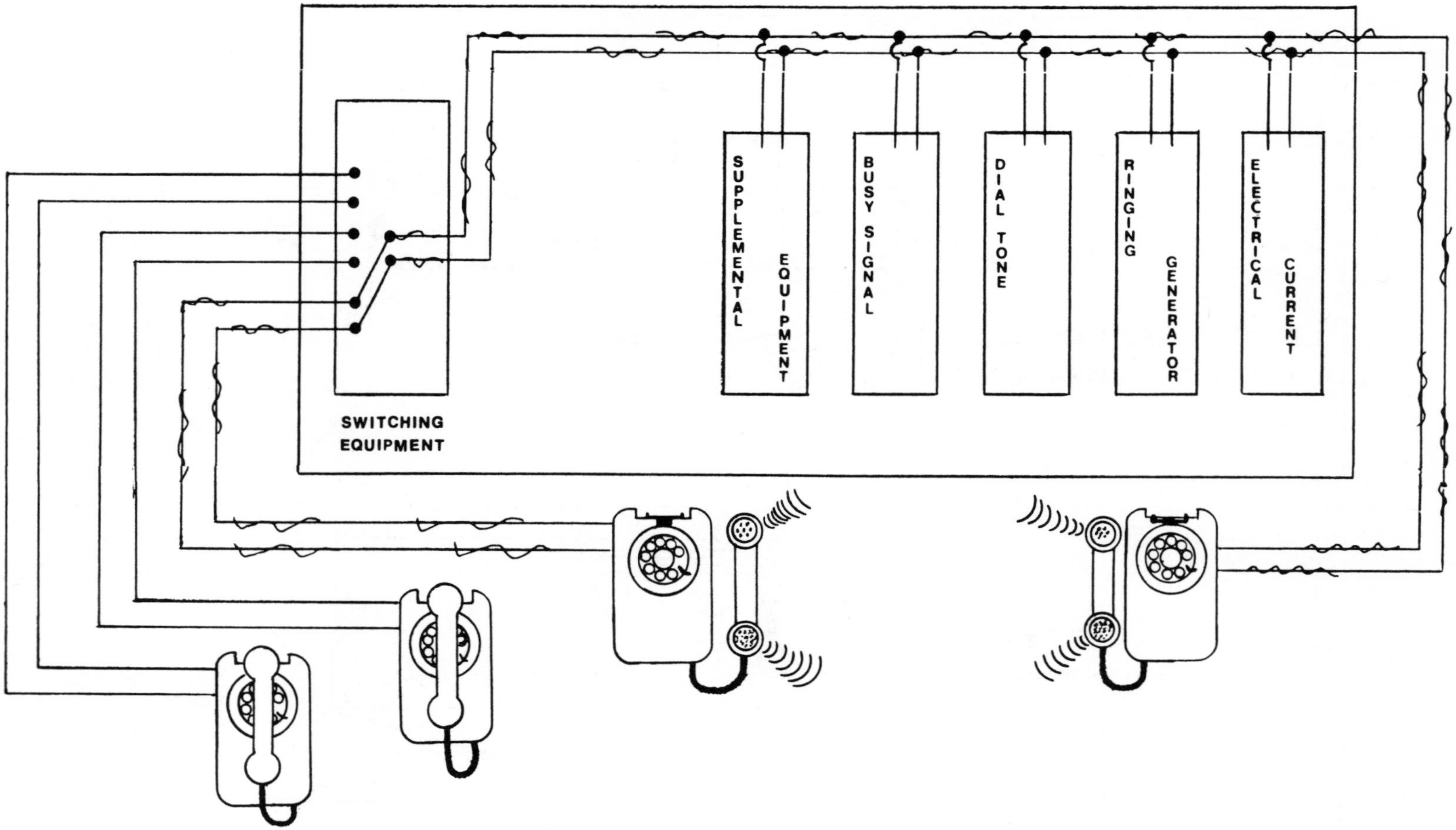

Fig. I-3 The central office.

are both found within a local network. Main cable is so named because it is the first cable to leave the central office and travel toward your home telephone. If your telephone is connected directly to the central office by the main cable, the connection is referred to as "main fed." Feeder cable originates from a *cross-connect terminal* (i.e., a terminating location for main cable and a starting location for feeder cable) and travels toward your telephone. A connection that uses feeder cable is known as "feeder fed."

Two or more cables often have to be spliced in order to increase a cable's length, change its direction, add a service or cross-connect terminal, or repair damage.

Service Terminals

The *service terminal,* located close to your home, is the final connecting point between the central office and your telephone. There are two types of service terminals; the type you have will depend on the type of home you live in.

The *protected service terminal* is found at multiunit buildings, structures in which individual units share walls, ceilings, and floors with other units (e.g., apartments, townhouses, and condominiums). This type of service terminal protects your telephone system against high voltages by providing fuses and a ground to earth for each pair connected to it. Figure I-4 illustrates one type of protected service terminal.

The *unprotected service terminal* is found at single-unit buildings, structures that are independent of others. The unprotected service terminal is used in conjunction with a house protector, which provides the necessary protection against high voltages. Figures I-5 and I-6 show two common types of unprotected service terminals.

Station Wiring

Your home telephone is linked to the local network through *station wiring,* which comprises all the wires and connections in and around your home that link your telephone with the service terminal (Figure I-7).

The arrangements for station wiring depends on what type of home you live in. Station wiring for single-unit homes includes the drop wire, house protector, inside wire, and phone jacks or phone modular connectors (PMCs). For multiunit buildings, the station wiring includes only the inside wires and the PMCs. Occasionally, single-dwelling station wiring will be found at a multidwelling location when there are only two or four units placed together.

Drop Wire

Drop wire connects the unprotected service terminal with the house protector. There are two types of drop wire: aerial and buried. Your drop wire will generally be the same type as your telephone cables. Aerial drop wire consists of one pair of wires coated with a heavy black rubber material. Locating aerial drop wire is easy, since it is usually in plain view,

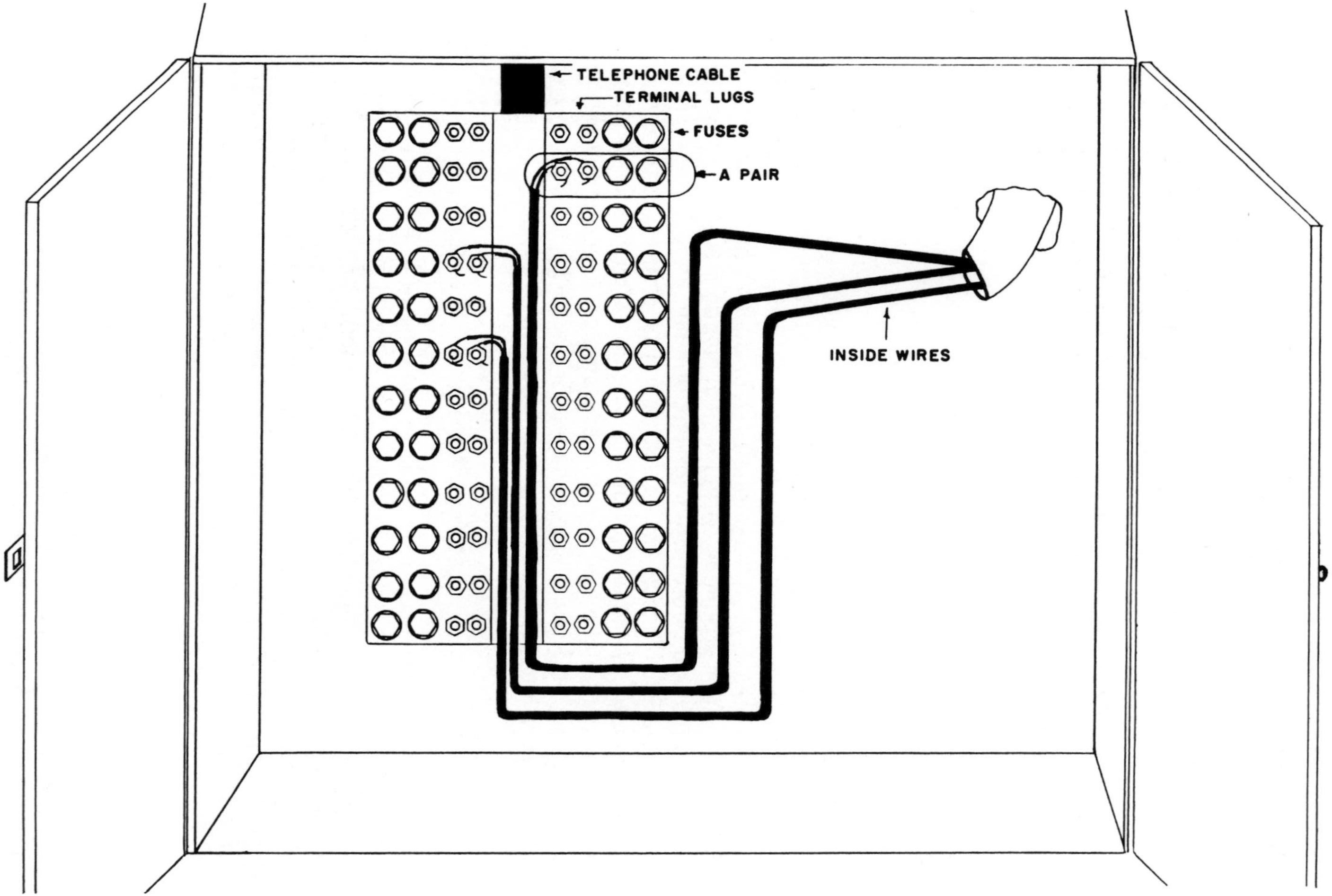

Fig. I-4 Protected service terminal.

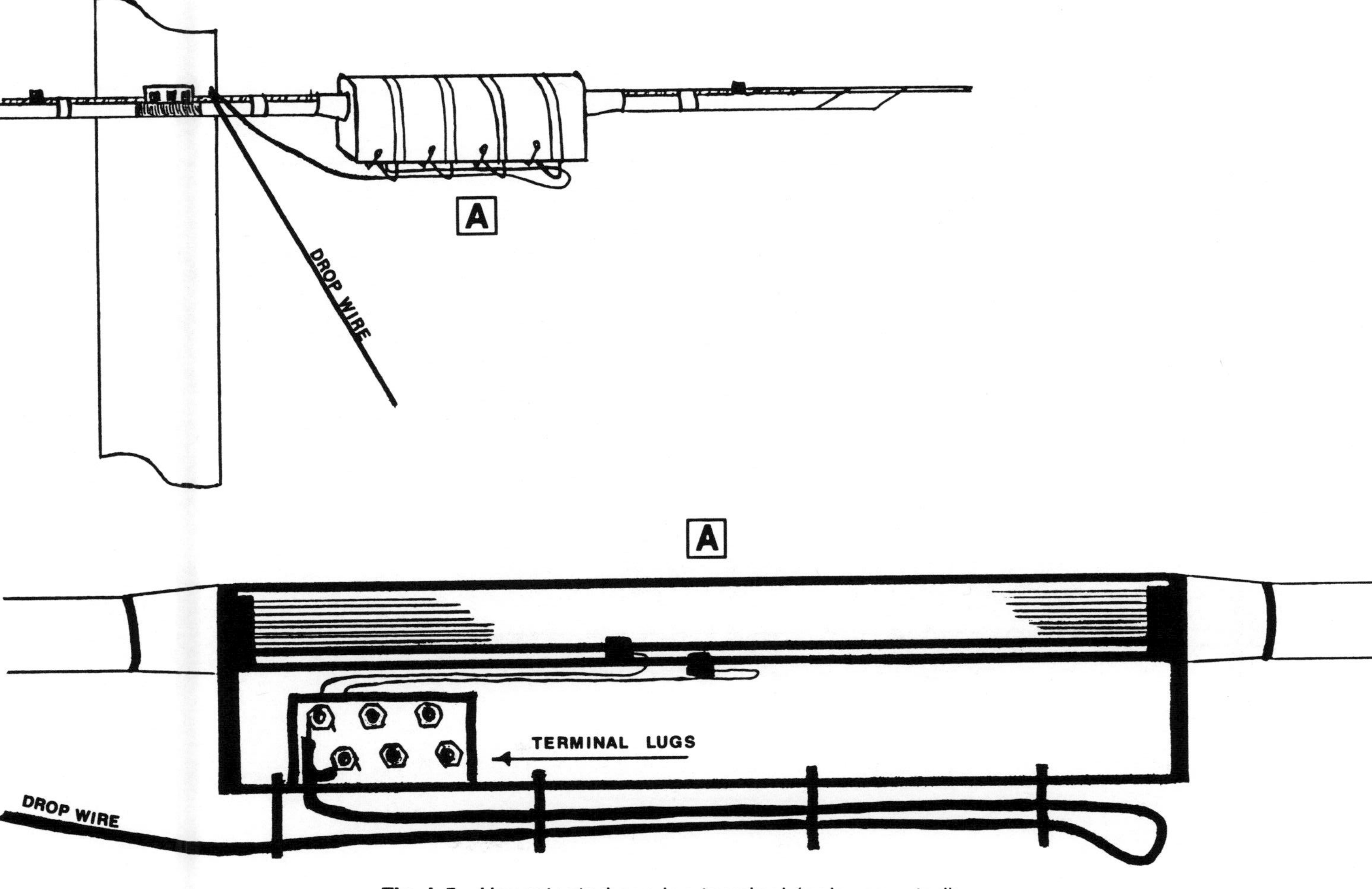

Fig. I-5 Unprotected service terminal (pole mounted).

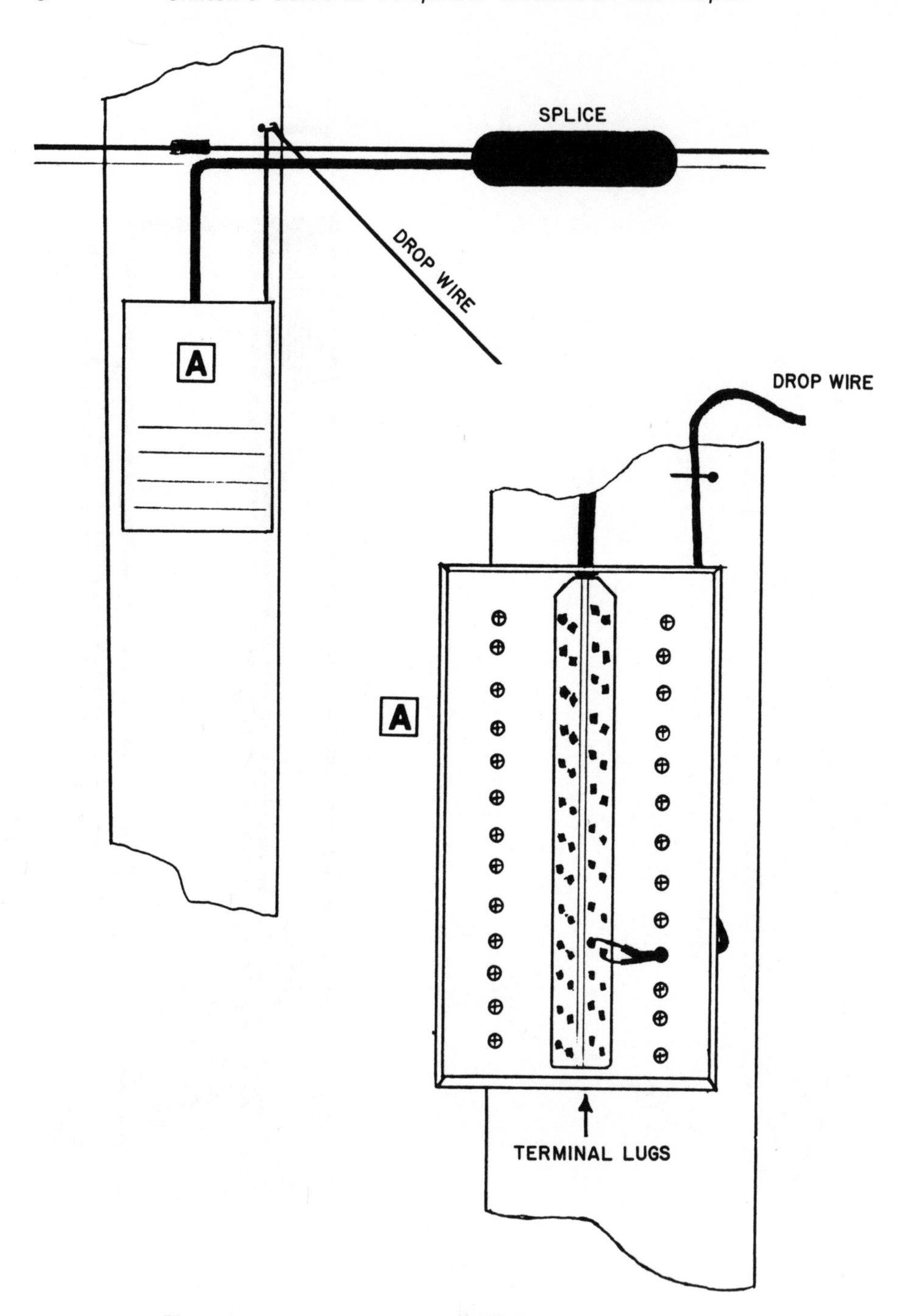

Fig. I-6 Unprotected service terminal (pole mounted).

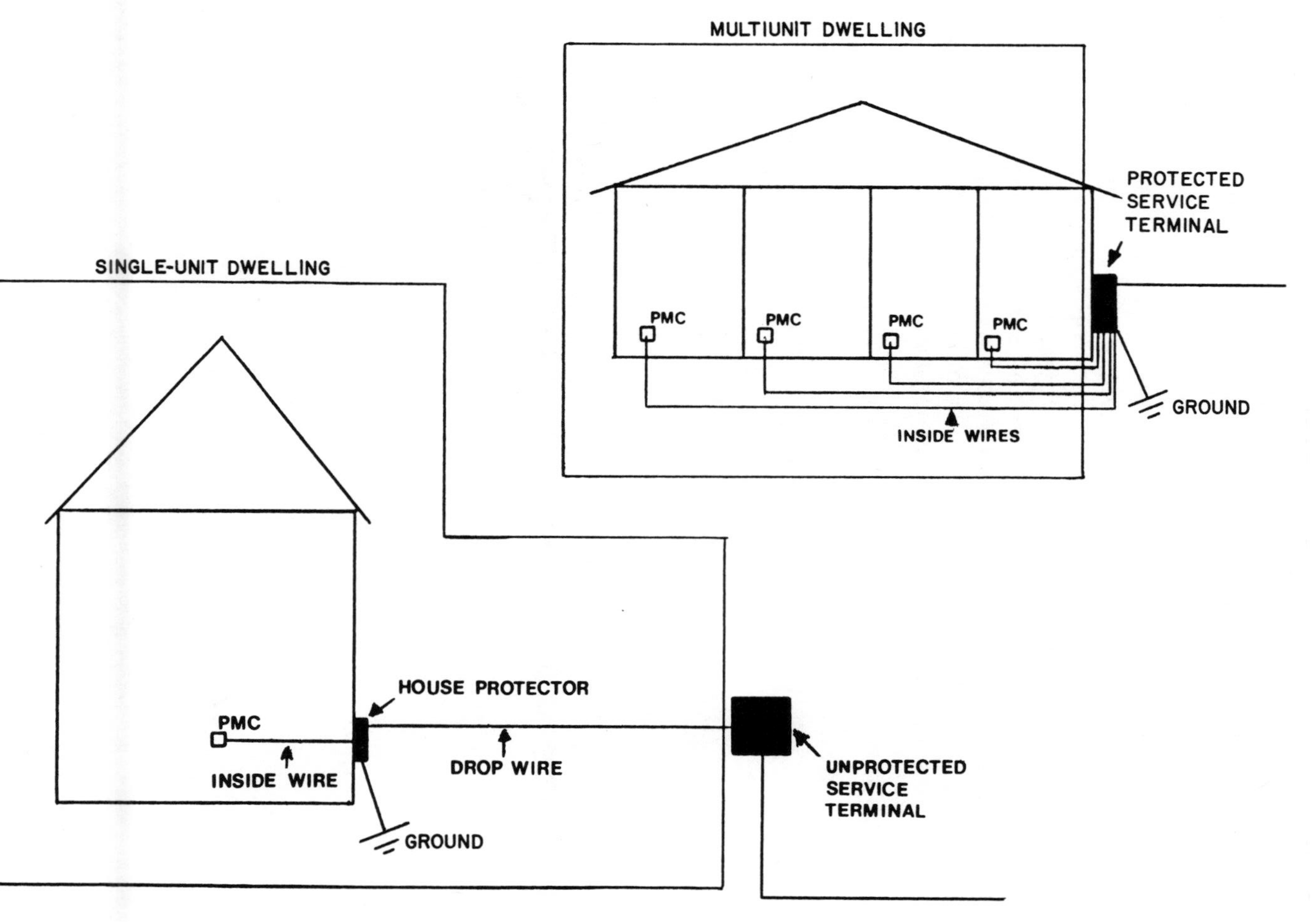

Fig. I-7 Station wiring.

spanning from the telephone pole to a leaning edge of the roof on your home (Figure I-8). Buried drop wire, in contrast, usually consists of three or more pairs of wires insulated by a black plastic material. Locating buried drop wire often requires assistance; Figure I-9 shows one possible location for buried drop wire.

The House Protector

The *house protector* is attached to the outside of your house, serving to protect it, your telephone, and you against high voltages. Fitted with fuses and an earth ground, just as is the protected service terminal, the house protector provides a location where all the inside wires can be attached. There are a number of different shapes, sizes, and types of house protectors, which will be covered in more detail in Chapter Three. Figure I-10 illustrates one common type of house protector.

Inside Wire

The *inside wire,* which connects the house protector or protected service terminal to the phone jack inside your home, may contain more

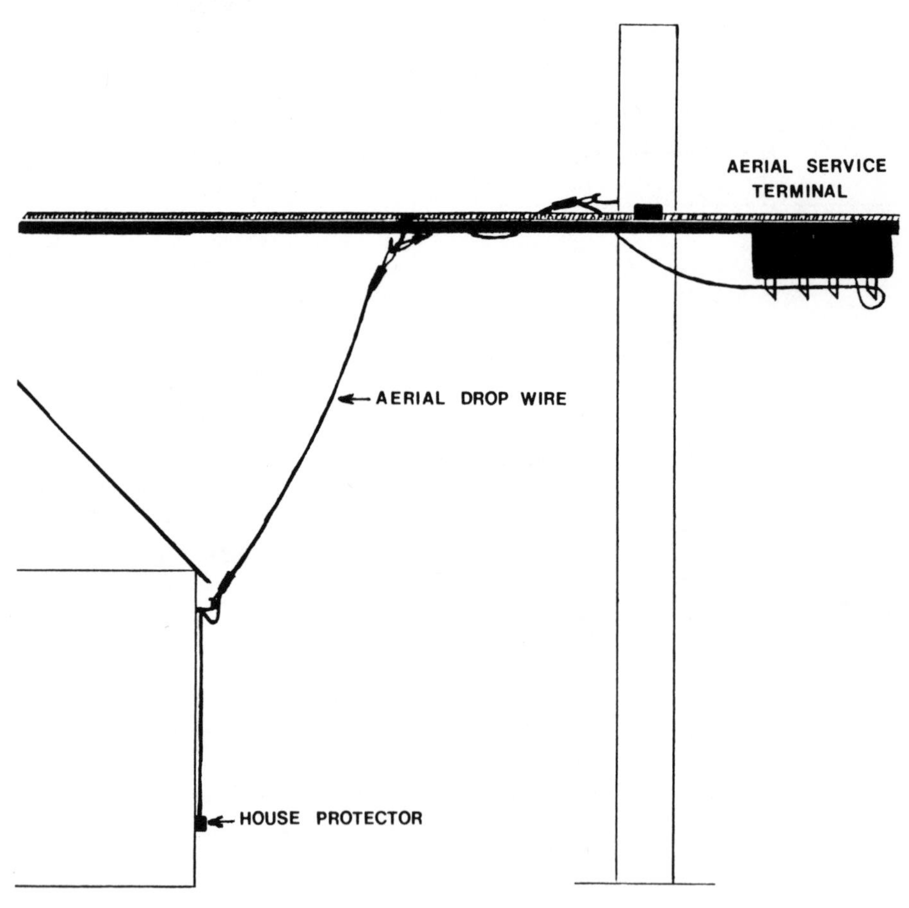

Fig. I-8 Aerial telephone cable.

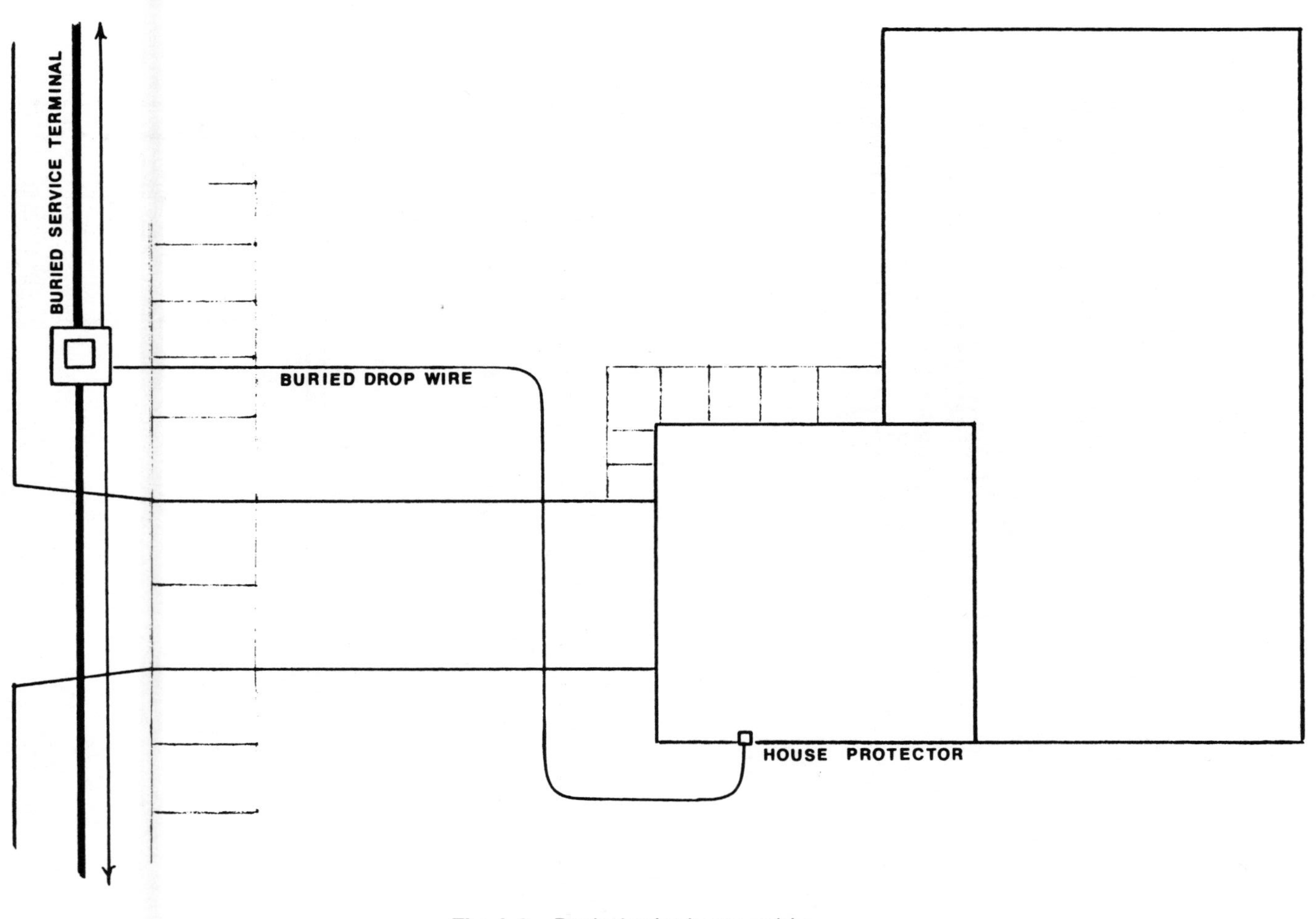

Fig. I-9 Buried telephone cable.

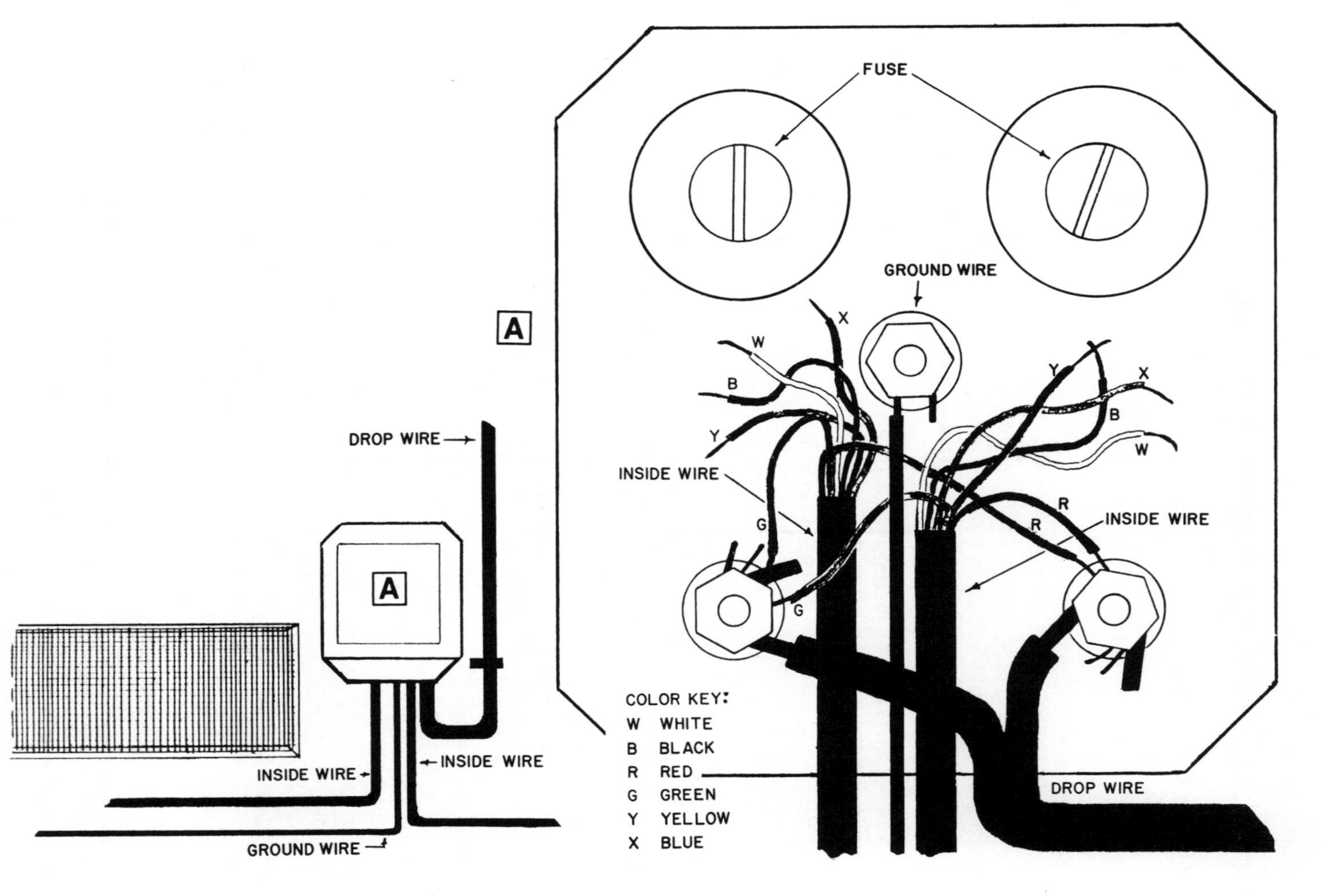

Fig. I-10 All-weather house protector.

than one pair of wires—some contain three pairs or more. Figure I-11 gives you one example of inside wire and where it may be found (there are different types of and locations for inside wire).

The Phone Jack—Phone Modular Connector

The *phone jack,* the point where you plug your telephone into the telephone system, is a terminal at which the inside wire is connected.

The *phone modular connector* (PMC) replaces all other types of phone jacks; it is now an accepted national telephone standard. Figure I-12 shows a variety of standard and nonstandard phone jacks.

Although you now have the basic information about how your home telephone system works, this introduction would not be complete without mention of the people who build, operate, and maintain the telephone systems.

Telephone Service Area

Your local network is joined with four or five other local networks to form a *telephone service area.* Figure I-13 maps out what a typical service area might look like as it links five central offices.

The telephone service area is operated by your local telephone company and includes three main departments.

The *business office,* responsible for the overall management of all departments within the telephone service area, handles budget allocations, accounting, general operating policies, local advertising, and customer billing.

The *customer service center* consists of two divisions: residential and business.

The residential department, responsible for receiving and scheduling installation orders for residential customers, provides price information and advice about available services within your local network. The payment of telephone bills will be accepted here, although usually the business office is responsible for billing. Recently, phone stores have been added to customer service centers, providing telephones, jacks, and wires for sale along with repair services for company telephones.

The business department provides similar services (with the exception of the phone store) to business customers. Because of competition among business telephone systems, this department provides an outside sales force as well.

Special crews in the *operation department* build, maintain, and operate the local telephone networks within their service area.

Telephone Installation Charges

Installation charges may vary from as low as $25 to as high as $150 or more within the same neighborhood.

Charges for installation are lowest when there is no need to route an installer into the field to physically connect your pair. This type of installa-

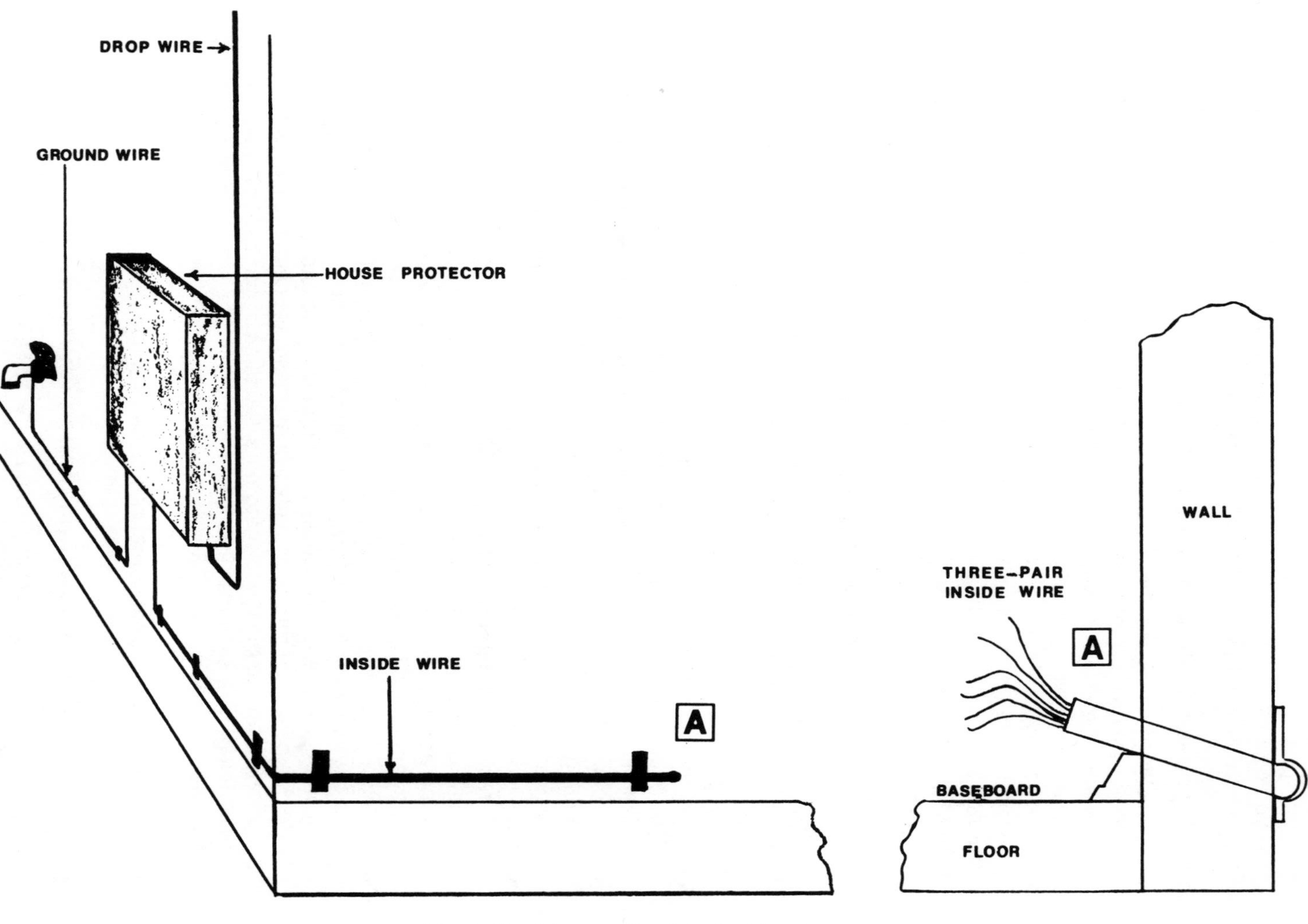

Fig. I-11 Wrap method for inside wire.

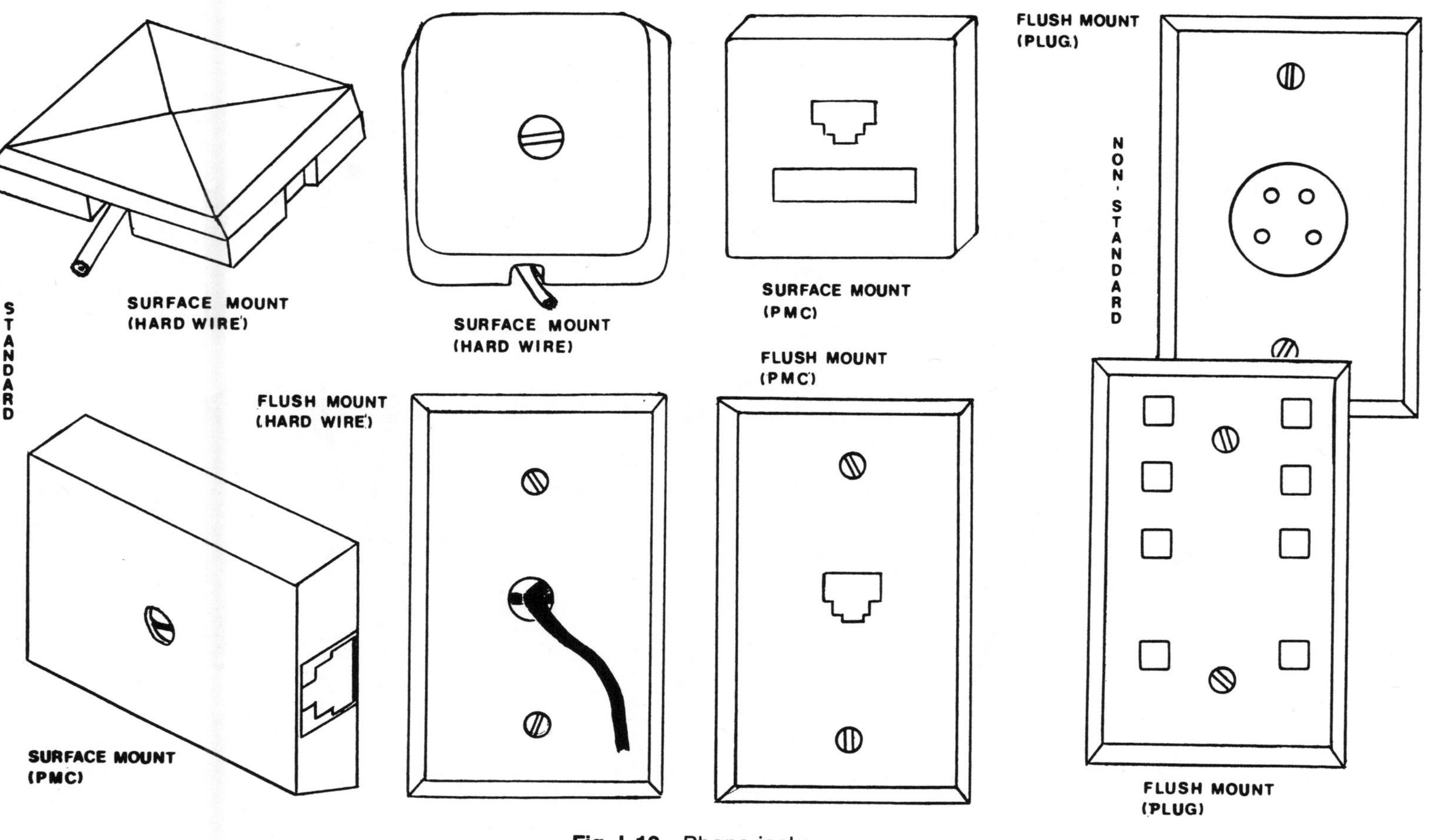

Fig. I-12 Phone jacks.

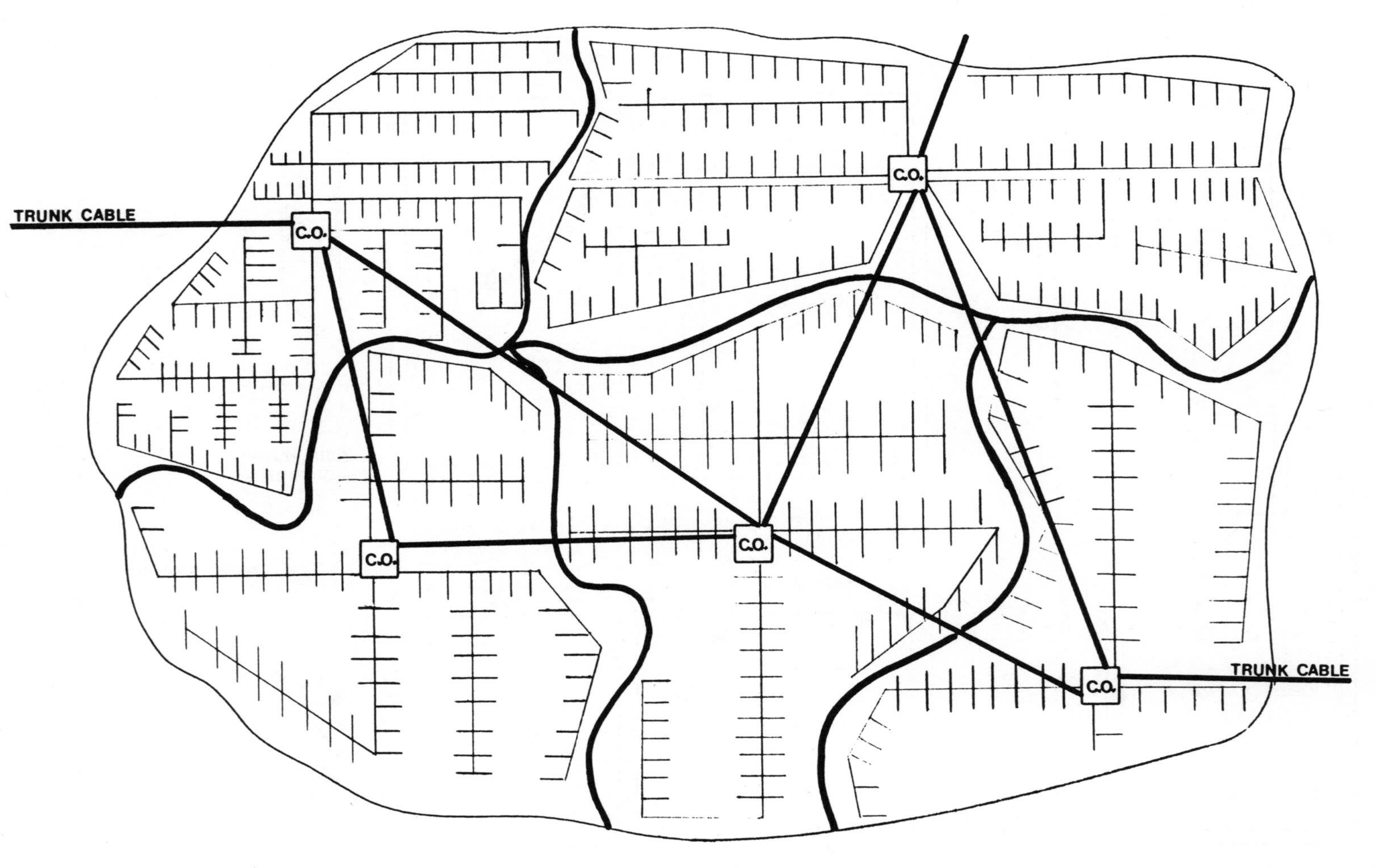

Fig. I-13 Telephone service area (with five local networks).

tion, called a home-run connection, is often accomplished by the telephone company at its central office.

A slightly higher fee may be charged for a field connect, where an installer physically connects your pair at a cross-connect terminal and/or a service terminal. There is an additional service visit charge if the installer is required to visit your home as well as charges for installing phone jacks and inside wires. Repeat service visits simply increase installation charges, so it pays to plan ahead.

The following figures are approximate charges for the various types of installations: home-run, $25; field connect, $25; service visit, $25; and installation, $25 per phone jack.

CHAPTER ONE

Telephone Instruments

This chapter begins by exploring the basic components of conventional telephone instruments (as opposed to the more contemporary instruments that are molded shut and therefore nonenterable, making component repair and replacement impossible). In addition, buying guidelines are offered to help you select a telephone suitable for your needs. Price ranges and special phone options, such as cordless telephones and memory dialing, are examined. For the budget-conscious consumer, we will discuss the advantages of buying versus renting a telephone.

Basic Components of a Telephone Instrument

The basic external components of the telephone are illustrated in Figure 1-1. This figure shows two types of phones: the desk rotary and the wall touch tone. Both can be easily entered for installation, component replacement, or repair.

The handset cord, consisting of four wires that connect the transmitter and receiver to the network, is usually the first component that will need replacing. On frequently used telephones, the cord often becomes stretched out, kinked, or even broken. Because most telephones come with short handset cords, many consumers install longer cords for more flexibility. (Make sure you purchase a telephone with a cord that can be unplugged at the instrument end; otherwise, if you have problems with the cord, you will have to replace the entire telephone instrument.)

There are three types of handset cords found on telephone instruments. If the *hard-wire handset cord* that is molded into the handset and base goes bad, there is nothing you can do except send it to the manufacturer or independent phone store for repair. With the *modular handset cord* (Figure 1-2), replacement is done simply by unplugging the old cord and plugging in the new. (*Note:* Modular handset cords have smaller plugs than modular line cords and are not interchangeable; also, some trimline

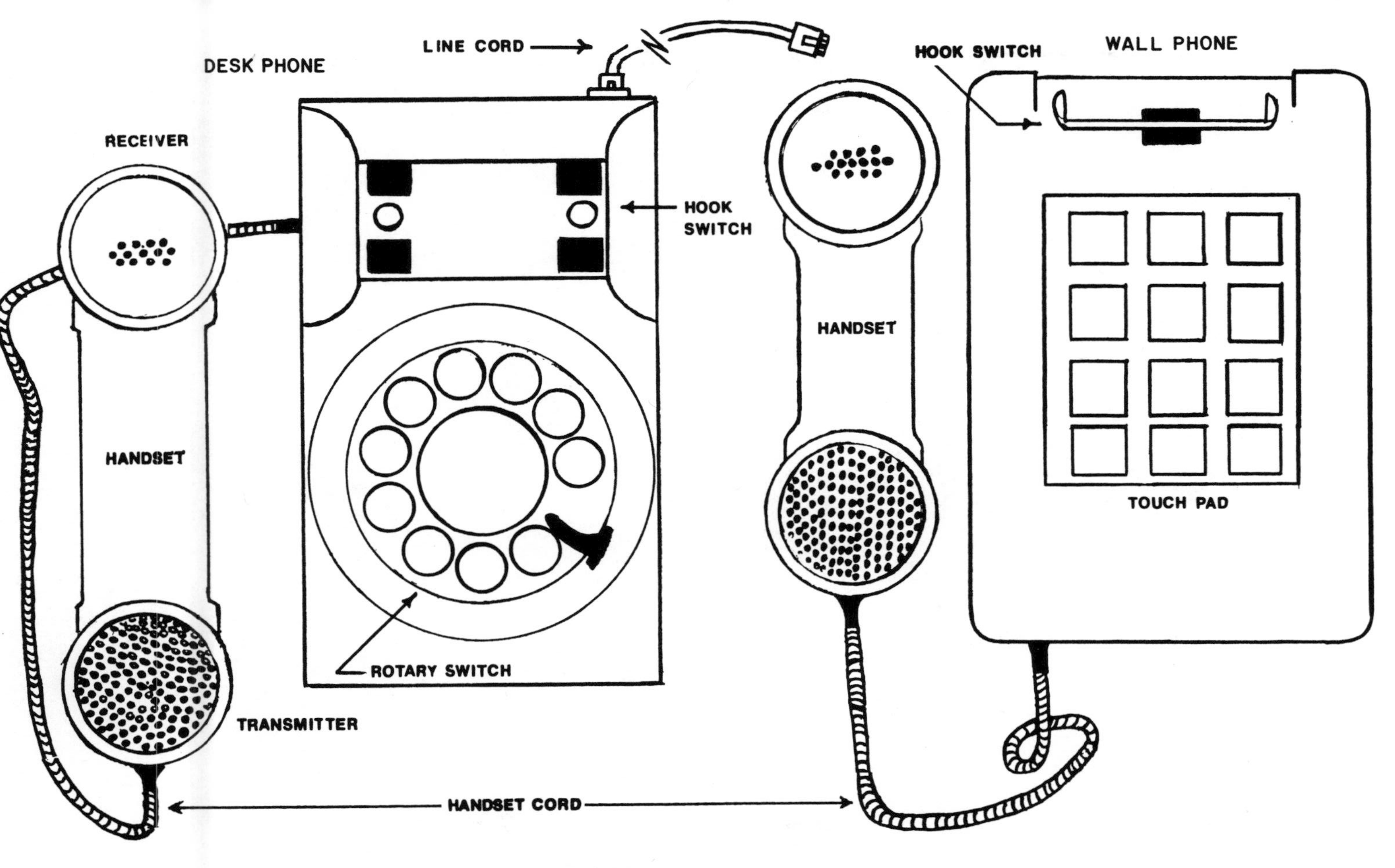

Fig. 1-1 External telephone components.

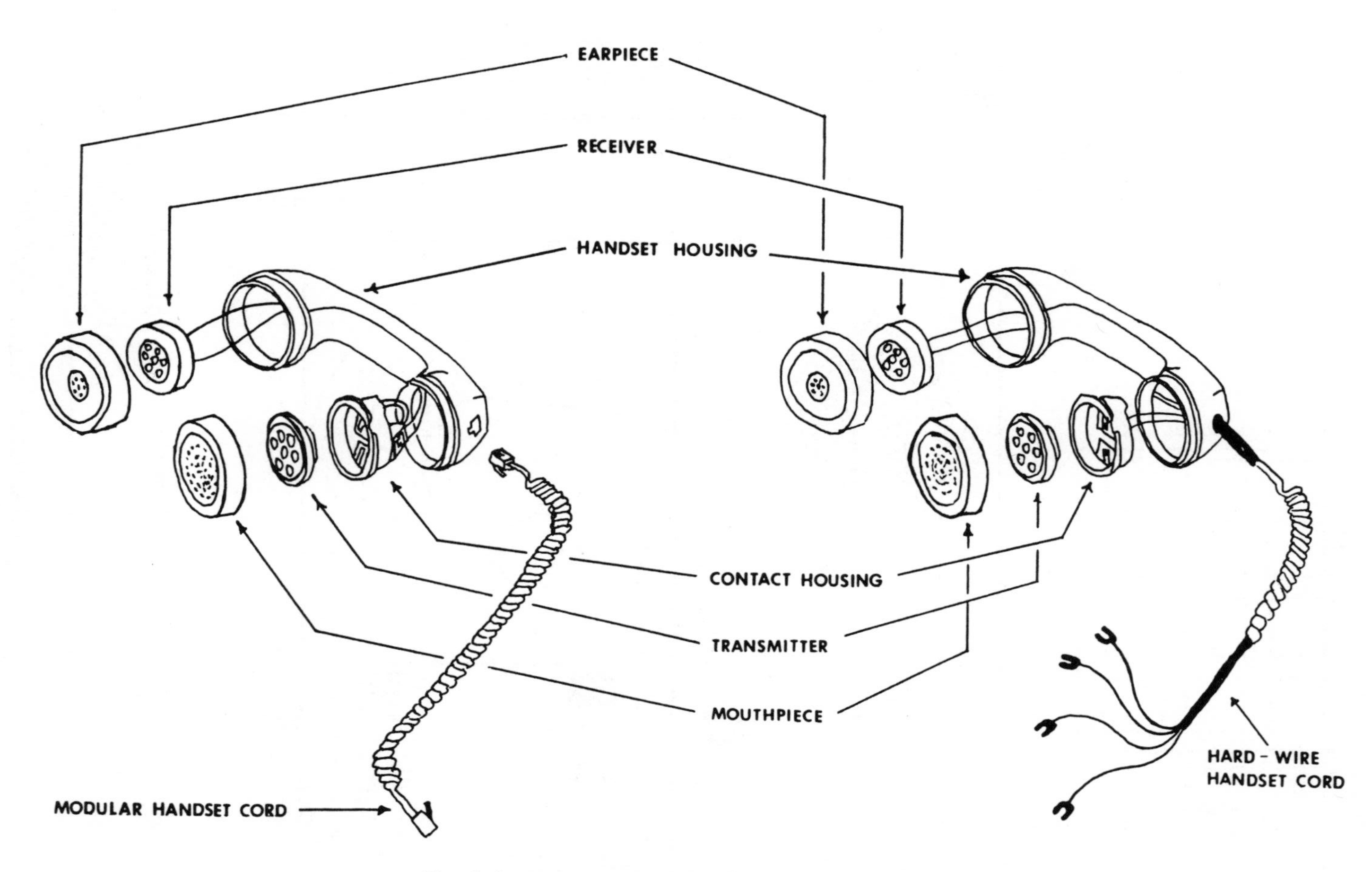

Fig. 1-2 Modular and hard-wire handset cords.

telephone instruments have larger handset cord plugs. Care must be taken when purchasing a replacement.) Figure 1-2 also shows the connection for the *hard-wire handset cord,* which is similar to the hard-wire line cord but has spaded ends on both the handset and network sides. Hard-wire handset cords are specifically designed to fit snug in the handset.

To replace the *transmitter* or *receiver,* simply unscrew the coverplate. Independent phone stores and local telephone company stores carry replacement transmitters and receivers.

To examine the internal components of your telephone instrument, you must first be able to get inside. Figure 1-3 shows how to open three common types. For the desk telephone, simply turn the instrument over. Unscrew the two screws that hold the cover on and then pull it off. For the wall telephone, insert a screwdriver in the small release tab at the base. The released cover will swing up and off. For the trimline telephone, remove the small rectangular plate and underneath you will find the screws that release the cover plate. Do the same for the handset to remove its cover plate.

Once the telephone instrument is open, the internal components will be exposed, as in Figure 1-4, which shows the inner workings of the desk and trimline telephone instruments. (The wall-mounted phone's internal components are similar to the desk phone's.)

The network is that point where line cords and handset cords connect. The easiest method for replacing these cords is to handle only one wire at a time: simply remove one wire from the old and replace it with the new.

If a telephone instrument is dropped hard, often the return spring attached to the hook switch will fall out of place and you will not be able to get a dial tone because the hook switch will not make proper contact. To repair this, simply open up the telephone instrument and reset the return spring.

Telephone instrument ringers are rated in a standard ringing power unit called the Ringer Equivalence Number (REN); every telephone should have an REN stamped on it. Telephone companies generally supply an REN of five to your home, so if your telephone instruments' RENs total more than five, your instruments' ringers may not work.

With rotary dialing telephones, sometimes you will hear the bell tap while dialing. You simply need to reverse the pair at either the phone modular connector or the network.

It is important for a telephone instrument to stay dry and clean. When dirt and grime build up on the touch pad (see Figure 1-1), they work their way into the contact switches inside, either shorting out the contacts or making the buttons sticky, which causes misdialing.

Surprisingly, cockroaches are a common maintenance problem. If they take up residence in your telephone instrument, they'll short out the wiring and jam up the switches. A periodic application of insect repellent (just a light mist) inside your telephone instrument should take care of cockroaches.

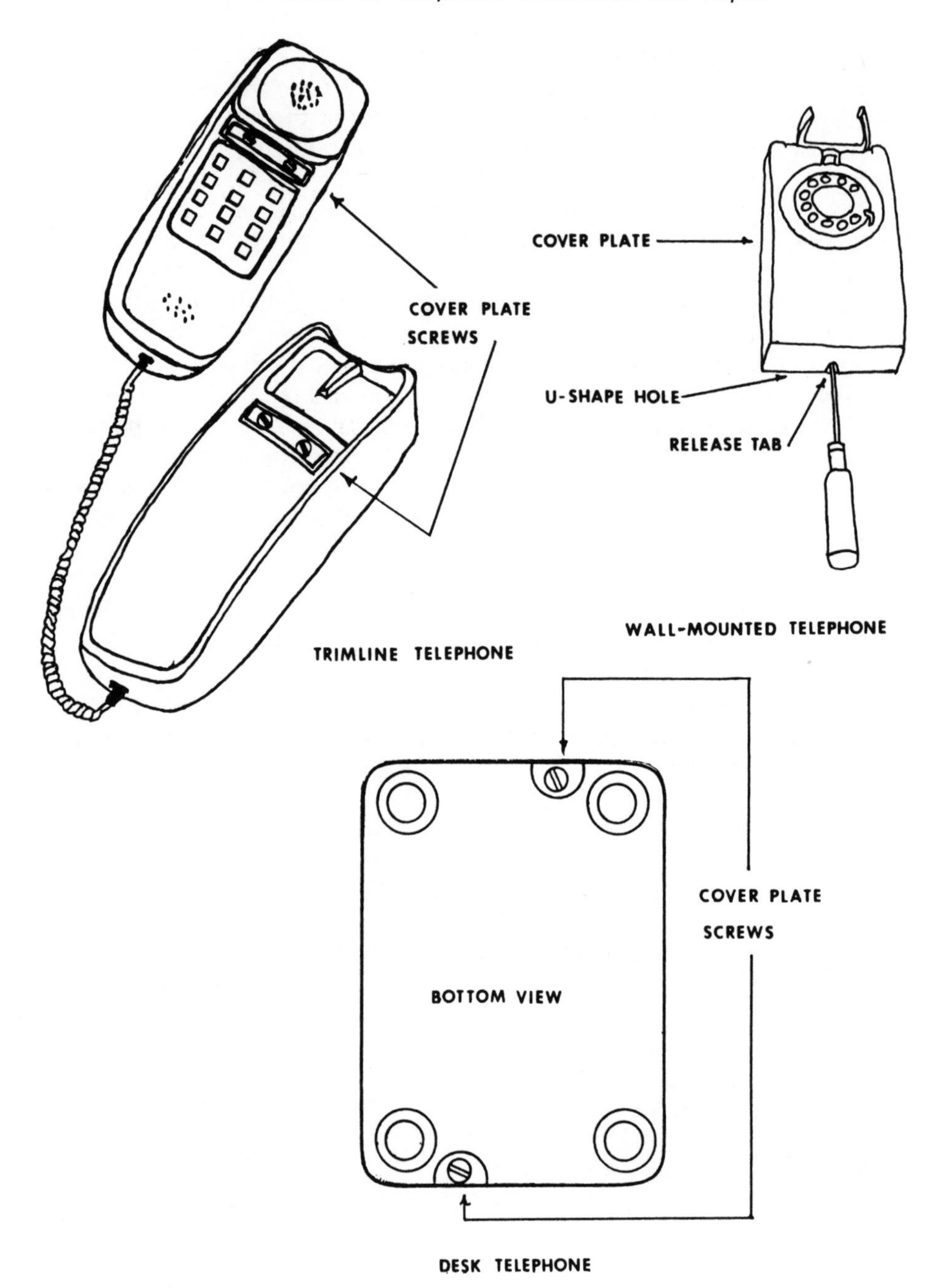

Fig. 1-3 Removing telephone instrument cover plates.

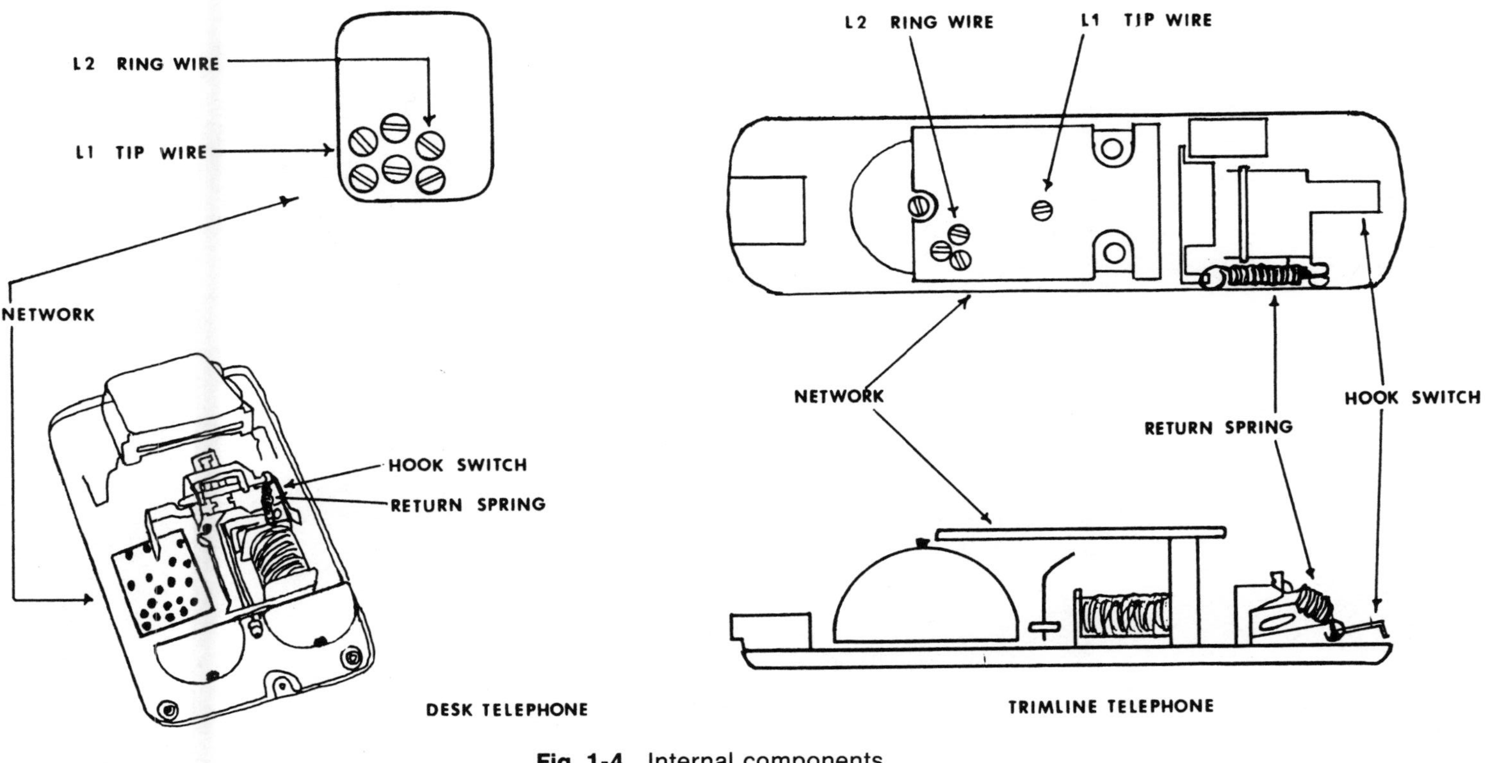

Fig. 1-4 Internal components.

Telephone Instrument Buying Guidelines

With such a wide variety of telephone instruments available on the market today, choosing the one that best suits your needs can be difficult. In addition to price, you need to consider a telephone's components, construction, and accessories. And always keep in mind the desired end result—communication. If a phone's sound quality is poor or static interferes with important calls, you won't be satisfied, even if you did save a few dollars.

Price Ranges

Telephone prices range from $10 to $200, with a one-piece, molded, pulse-dialing telephone with no options at the low end and a sophisticated, custom-design telephone loaded with options at the high end. Regardless of the price, all telephones will have been approved by the Federal Communications Commission (FCC), which means they have been standardized to operate within the telephone network of the United States and are safe to use.

$10 to $30 Telephones

At standard retail prices, telephones within the $10 to $30 price range are generally the molded, one-piece telephones with pulse dialing and few if any options. They often have poor voice transmission quality, and because the hook switch is probably integrated within the handset, next to the transmitter, your chin may depress the switch while you're talking, accidentally disconnecting your call (see Figure 1-5). These low-priced phones are typically light in weight (less than half a pound); therefore, it is sometimes difficult to hang them up properly, and if dropped, they will more than likely become damaged beyond repair. When line cords or handset cords go bad, replacement is impossible because they are molded; thus, the telephone becomes worthless.

Options included with these types of telephones are generally limited to a memory capacity of only two or three telephone numbers and maybe a ringer control consisting of "on" and "off." Many of these telephones claim to operate on tone dialing and rotary dialing lines; however, when the phone operates on tone dialing lines, pulse dialing, not Dual Tone Modulating Frequencies (DTMF) is incorporated. DTMF is preferred to pulse dialing because it is much faster and is required to access long-distance discount dialing services. In addition, some answering machines and other services, such as automatic telephone banking and traveling long-distance dialing codes, require the DTMF signal. If you are already paying for DTMF dialing lines, it doesn't make sense to use pulse dialing on them.

Warranties and service for the $10 to $30 telephones are typically the worst in the telephone instrument industry. Warranties usually last for only 30 to 90 days, and service is usually handled through the mail, taking, in some cases, up to six months.

Although I do not recommend the $10 to $30 telephones for general

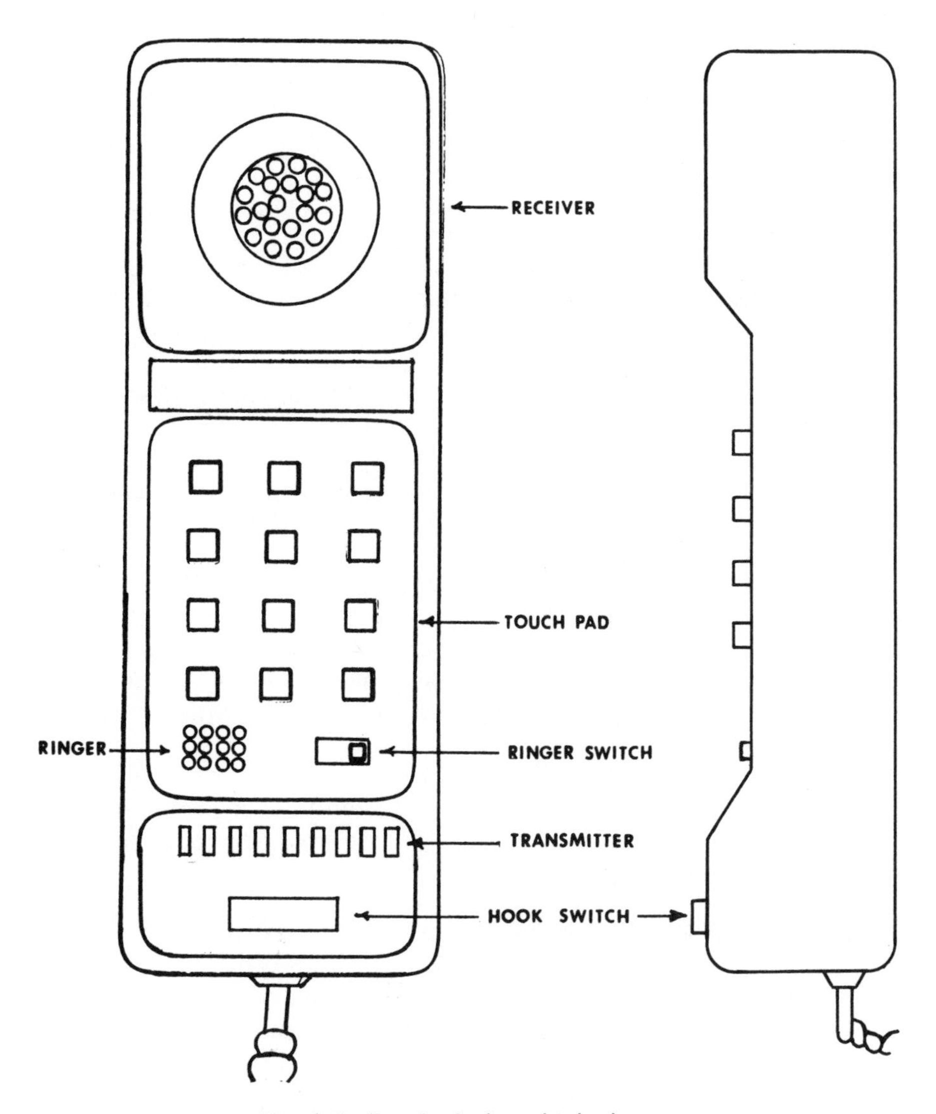

Fig. 1-5 Poorly designed telephone.

use, they are effective as test instruments or standbys in emergencies since they are inexpensive and can be stored easily on a shelf.

$30 to $60 Telephones

It is possible to purchase a reliable telephone with quality voice transmission, solid operational characteristics, and a few useful options within the $30 to $60 price range.

Stores such as Best, Target, Sears, and K-Mart often offer special buys on these phones. At the upper end of this price range, you'll find the basic quality telephones with such well-known names as AT&T, GTE, Bell, and Panasonic. These come in the three most popular models—desk, trimline, and wall-mounted—and offer a number of options, such as ringer volume control, two or three telephone number memory capacity, lighted dial or touch pad, redial, mute button, and DTMF tone or rotary dialing. These phones also come with modular line and handset cords for easy replacement.

Another way to buy a $60 telephone for less than $30 is to check out garage sales, thrift stores, and swap meets. You may be able to pick up a brand name telephone needing only a minor repair for $10 to $20. Another $5 to replace the line cord and handset cord or repair the hook switch return spring may be necessary, but you've just bought yourself a quality telephone.

Beware of telephones selling near $60 with unknown brand names offering many telephone- and non-telephone-related options, such as radios, clocks, and intercoms. At this price range, quality is usually sacrificed for quantity to lure you into a purchase. For quality telephones offering many nonrelated options, you need to look in the $100 and up price range.

Some unfamiliar brand names do offer good deals at lower prices. I would suggest a store such as K-Mart, which carries quality product lines and will take back products that do not satisfy its customers, to test lesser known brand names. Also, K-Mart's telephone center is stocked with a good line of telephone installation, repair, and conversion kit materials.

$60 to $100 Telephones

You will be able to purchase a quality brand name telephone with a few extra options or a quality telephone with nonrelated telephone options within the $60 to $100 price range.

$100 to $200 Telephones

A wide variety of quality brand name telephones with a combination of nonrelated and related telephone options is available within the $100 to $200 price range. Telephones at these prices offer two-line capability and hold buttons allowing the user to transfer from one line to the other, similar to business office key system telephones; quality speaker telephones; and telephone answering machine combinations. The custom-design telephone is also found within this price range. These are the telephones cased in decorator shapes, ranging from footballs to executive woodgrain boxes.

Construction Styles

The basic construction of a telephone instrument needs to be solid to ensure against everyday knocks and thumps. If the phone tips over easily and goes off the hook, you may miss incoming calls. In addition, you must assess construction in terms of comfort and convenience of use. Is there any chance of accidentally disconnecting the call with your chin? Can you comfortably cradle the receiver between your head and shoulder?

The most important point to consider is the phone's ability to withstand a sudden, sharp impact. If a telephone cannot withstand a fall to the floor, you'll end up paying substantial repair bills or possibly having to throw it in the trash. Telephones that are constructed of a heavy-duty plastic and that weigh between three and four pounds usually stand up under this type of impact because the heavy-duty plastic offers an absorption factor that the lightweight plastics do not. The heavy-duty components inside, such as the ringer, network, dial, or touch pad, are also able to handle sudden impacts.

If you drop your phone, give it a gentle shake to ensure that none of the inner components are loose. Loose components create static and cause broken wiring, which makes the telephone inoperative.

The ideal telephone will have a construction that allows entry into its components for replacement, additions, and alterations. As mentioned earlier, you will want to ensure that new cords can be added to the handset. Entry is also vital for such alterations as wiring the telephone for a locking device that disconnects the rotary dial or touch pad by key. In some telephone service areas where the central office equipment is older, the ringer component must be connected to the telephone circuit in a special way that requires entry into the instrument. Also, if you are purchasing a telephone for a party line telephone number, the ringer circuit of the telephone will require modification, which means the telephone must provide access to its inner components.

Unfortunately, many of the newer telephones on the market today lack the proper human engineering construction (i.e., efficiency and comfort of use). On phones where the hook switch is located right next to the transmitter in the handset (see Figure 1-5), you must pay special attention to where you put your chin while talking to avoid disconnecting the line. As a result, you may have to keep your mouth at a distance from the receiver, which causes the listener to have to strain to hear.

Another problem is the telephone that has option buttons located within the handset, which requires the user to hold the handset by the tips of the fingers to avoid activating an option by mistake. This becomes an uncomfortable position for the hand, especially when conversations last for more than an hour.

Comfort also comes into play when a caller enjoys talking on the phone while writing messages, working on needlepoint, washing the dishes, or dressing the baby. Some telephones simply won't allow you to cradle the handset between your shoulder and head. You will need to decide if this is an important factor before you make your purchasing decisions.

Options

With the purchase of a telephone comes the chance to select any of a number of options. The following is a list of the most common options now available.

Rotary Dialing

Rotary dialing was one of the first big steps in offering the home telephone user independence from operators, who used to have to manually connect telephone circuits.

By turning current off or on in the local-network cable pair, rotary dialing provides the telephone system with the telephone number. Switching equipment at the central office interprets the current and connects the call's outgoing pair. Pulse dialing, similar to rotary, does this electronically. (Rotary or pulse dialing telephone instruments will work on touch tone circuits, but touch tone telephone instruments will not work on rotary dialing lines.)

If your rotary dial is broken, it is possible to dial a telephone number by depressing the hook switch in a Morse code manner. For example, the telephone number 853-1212 would be dialed by depressing the hook switch eight times, pausing, then depressing the hook switch five times, pausing, then three times, pausing, and so on until the number is completed.

Since rotary dialing circuits are not as expensive as touch tone circuits, the consumer may be able to save money; however, many types of computer and long-distance discount dialing services are not accessible with rotary dialing.

Touch Tone Dialing

Touch tone dialing sends combination tones, called Dual Tone Modulating Frequencies (DTMFs), to the central office.

This is the type of dialing you will need for many computer and long-distance discount dialing services. Although rotary telephone instruments will work on touch tone circuits, touch tone dialing is approximately ten times faster than rotary.

Mute

The mute option performs a function similar to that of cupping your hand over the receiver to prevent background noise at your end from being heard by the listener. The mute button disconnects the circuit inside the telephone that connects the receiver and, therefore, blocks all noise from being heard by the listener. A built-in feature, the mute option is useful if you need to pause in your conversation to give instructions or ask questions of someone else and you don't want the listener to overhear; however, he or she will be aware that you have activated the mute feature because the noise level will suddenly become completely silent.

An add-on feature similar to the mute option but that eliminates only background noise and does not disconnect the receiver circuit is called an external noise limiter, which is exchanged with the mouthpiece unit of the handset. With it, the listener will not be aware of any noise level changes and, therefore, will not be aware that you may be talking with someone else. To add the external noise limiter to a telephone, you will need to have an instrument with a modular handset, as shown in Figure 1-2.

Ringer Control

A telephone without ringer control is almost like a stereo without volume control; you're stuck with a fixed volume. If your phone's ring is too soft you may miss important calls and if it's too loud you may jump out of your shoes.

Some telephone instruments come with no volume adjustment but offer the option of switching from ringer to flashing light. Many newer phones have a ringer that sounds like a cricket as opposed to the conventional bell sound.

Redial

This option allows for the automatic redialing of the last telephone number you dialed. This is a handy timesaver for telephone numbers that are busy, because you simply depress the *redial* button and the number is dialed again. Some telephones offer an extended redial feature that redials the last busy telephone number you dialed up to fifteen times, allowing you to go about your business while the telephone automatically does the work for you.

Memory Dialing

Memory dialing is an interesting accessory designed to facilitate telephone calling. It allows you to store telephone numbers and recall them simply by dialing one or two digits. Because it can save you up to 80 percent on dialing time, it is extremely useful in emergencies.

Generally, telephones with memory dialing come with three-number capability; however, some offer up to 60-number capability. Or you can purchase a separate component with memory dialing capability. Another way to add memory dialing to your home phone is to request speed dialing from your local telephone company. This added service provides a memory space in the switching equipment at the central office for telephone numbers you wish to add.

If you purchase a telephone with memory dialing capability, you should get one with a battery power backup system so that if the electrical power should fail, the numbers stored in the telephone's memory bank will not be lost. If you use a long-distance discount dialing service that requires an access and authorization code, a memory dialer with a pause feature will allow you to dial these codes in the proper sequence.

During an emergency, dialing a seven-digit number to contact the police, firefighters, paramedics, or poison control officials can be difficult. If emergency numbers are put into a phone's memory, contact with the necessary party can be accomplished simply by pressing one button or dialing one digit; therefore, your chances of misdialing are reduced and, as a result, the response time of the emergency department may be decreased.

By limiting your phone's memory to three or four numbers, you can ensure that a person not familiar with your telephone's operating characteristics will be able to use it in an emergency.

There is one drawback, however, with memory dialing. If you suddenly find yourself away from the phone that carries this option, and you need to reach a certain party, you may be stranded if you can't remember the phone number.

Hold

The hold option, which allows you to put your caller on hold, is similar to the Key or PBX telephone systems found in many business offices. This feature can be purchased either as a part of the telephone or as an add-on option. Make sure that the hold option you're considering buying doesn't have a two- or three-minute time limit. In the home, the hold option allows you to put a caller on hold, hang up that particular telephone, and go to a more convenient phone to resume the call.

Speaker Phone

The speaker phone feature is great for the kitchen, baby's room, or home office because it allows you to continue a telephone conversation while going about your business. With this option, both your hands are free and you can walk around the room since the handset is left on the hook switch when the speaker phone is activated. A word of caution, however: some people do not like their conversations going through a speaker phone because any third party can overhear what may be a private conversation.

Tone/Pulse Dialing

Many new telephones have pulse dialing by push button, which is similar to rotary dialing and works on rotary and touch tone (DTMF) circuits; however, it will not work on long-distance discount dialing services or many computer services because they require DTMF signals. (Remember, touch tone telephone instruments do not work on rotary circuits.)

Since the tone/pulse dialing option allows the telephone to be dialed on either DTMF or rotary dialing telephone circuits, this feature is a good one if you frequently move and do not know what dialing circuits will be available to you.

Call Timer

The call timer option, not a common feature, times how long you spend on any particular call. It is handy for comparing the cost of long-distance calls with different telephone companies.

Number-Dialed Display

The number-dialed display option, another uncommon feature, displays on an LED readout the number you just dialed. It is helpful for checking that you haven't misdialed.

Lighted Dial or Touch Pads

The lighted dial or touch pad feature provides a small light to your dialing digits that is very useful for nighttime calling. Allowing for quick and accurate dialing without the need to turn on bedroom lights, this option permits you to place a call in the middle of the night without disturbing your sleeping partner.

In an emergency, the lighted dial or touch pad will enable you to call the police without turning on a light, which could alert intruders.

Adjustable Receiver Volume

The adjustable receiver volume option, which lets you turn the volume level up or down, is handy for people with hearing deficiencies or for calls that are low in volume. The adjustable receiver volume option can be purchased as an add-on feature; it requires a modular handset because the device is connected to the handset cord.

Multiline Telephones

A variety of multiline telephone instruments accept two and three telephone numbers at a time. Remember, it takes two wires (a pair) to make a telephone number. If you get the two-line telephone instrument, you will need to wire your phone modular connector as shown in Figure 3-22 in Chapter 3, and if you get a three-line instrument, you will need a six-pin phone modular connector as shown in Figure 3-21 in Chapter 3. Keep in mind that for three lines you need three pairs.

Cordless Telephones

The cordless telephone offers the consumer freedom from the restrictions of line and handset cords and serves as an alternative to installing phone modular connectors.

Figure 1-6 illustrates the basic components and operating ranges for cordless telephone instruments, which are made up of two units: the base unit and the handset. Within the base unit are a power cord that supplies power to the electronic components and to a battery charger inside the handset and a modular line cord for connecting the cordless telephone instrument to a phone modular connector.

Features include indicators to signify if the telephone is in use when the handset is away from the base unit and battery charge level indicators to let you know if the handset has a proper charge. Some base units even include built-in intercom systems that allow the phone to operate like a walkie-talkie.

The sophistication of the cordless telephone instrument determines which features are included, though most are similar to the ones found on the newer-type telephone instruments. (*Note:* Some handsets do not come with touch pads, which means your cordless telephone will only be able to receive calls.)

Operating on radio signals, the cordless telephone instrument is subject to the same type of interference that most radio signals experience. When you move around, the shifting of the antenna position will distort the receiving and transmitting signals, which means your conversations may not have the sound quality you would prefer. It is also possible that your cordless telephone may receive interference from ham or CB radios transmitting strong signals in close proximity to your location.

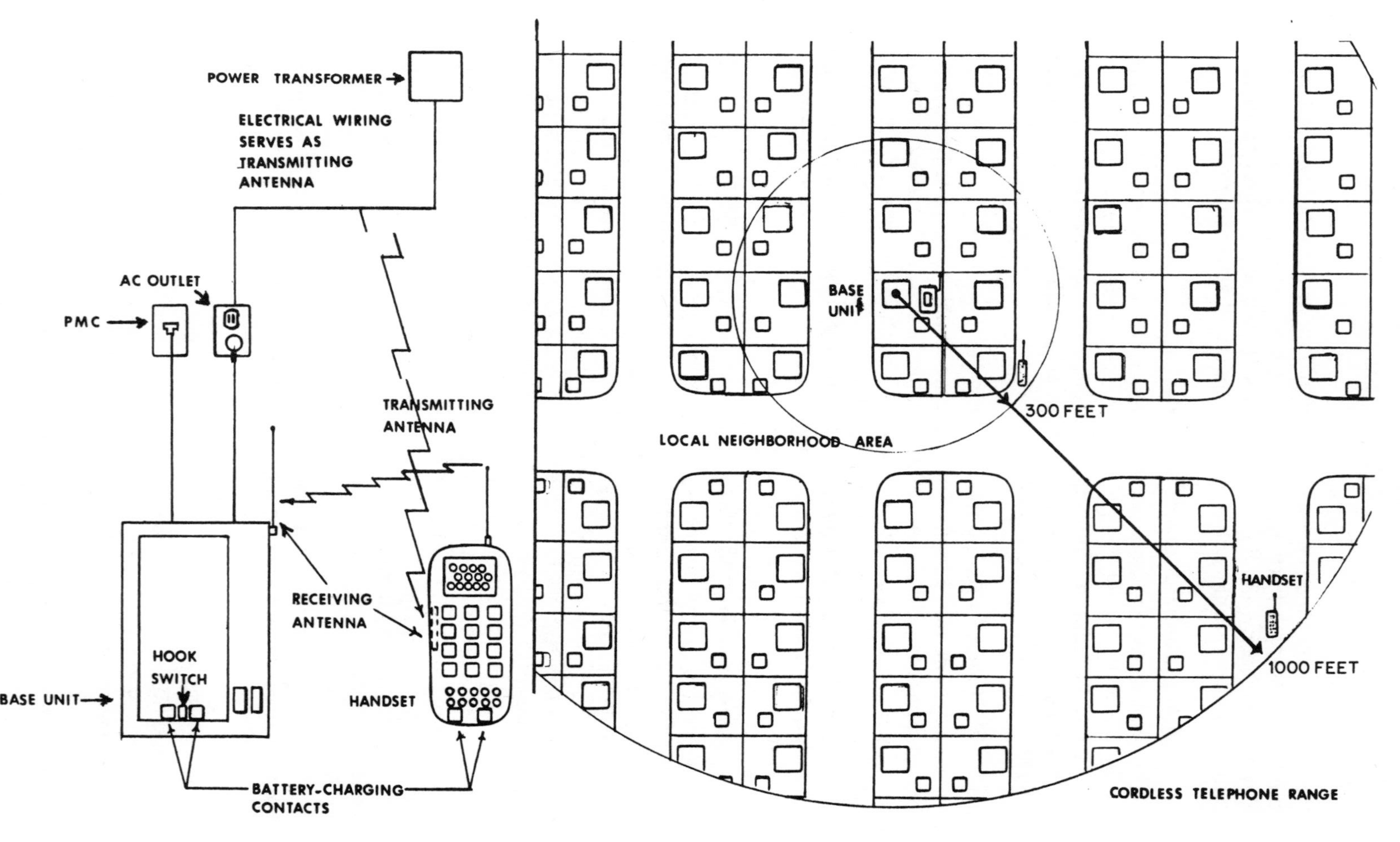

Fig. 1-6 Cordless telephone.

Interference from other cordless telephone instruments in your area means it is possible that other cordless phones can be receiving your conversations and vice versa. This situation can also result in false ringing, which is annoying. Although many manufacturers have added features with names like "security switch" or "private button" to eliminate false ringing, they cannot eliminate other cordless telephones from receiving your conversations when they are on identical frequencies.

The cordless telephone is not the instrument to use if your conversations require privacy and confidentiality. In fact, there is federal legislation pending on the issue of recording cordless telephone conversations and whether or not it is legal to do so without a court order. A number of lower state courts have already ruled that telephone conversations that take place over cordless telephones, using air waves, are in the public domain and therefore are not subject to any privacy laws that may protect conversations.

Not only is interference a common problem, but when two nearby cordless telephones are on the same frequency range, they may be able to access each other's telephone circuits to make outgoing calls. Under such circumstances, a certain amount of confusion is bound to occur over who made which calls. Generally, telephone company policy will be to investigate any possible billing errors, but it will not adjust or demand payment from one customer to settle a claim raised by another.

Cordless telephone customers can practically eliminate this problem by activating the "security switch" whenever the cordless telephone is not in use. If your phone doesn't have the "security switch" option, simply unplug the instrument's base unit from the phone modular connector.

For the best cordless telephone communications, purchase a phone with a range of 300 feet; the potential number of other cordless telephones within the 300-foot range will be a lot less than that in the 1,000-foot range (see Figure 1-6).

Warranty and Service

Important factors to consider when deciding which telephone to purchase are the warranty and service agreements that accompany each telephone.

With telephones in the lower price range, not much is guaranteed in the area of warranties and service. Generally, only 30 to 90 days are offered on parts and labor, and service is typically handled through the mail, which may take months.

Mid to higher priced phones offer a wider range of warranty and service agreements. Warranties can go from 30 days on parts and labor to up to three years, and service arrangements may include convenient local repair shops or mail-in services that offer a loaner telephone while yours is being repaired.

Buying Versus Renting

Not too long ago we all rented our phones from the local telephone company. Repair personnel would actually come into the home to repair problems with the station wiring or instrument at no additional charge. But those days are over. Whether you rent or own your phone, the telephone company will still send a serviceperson out to find the problem. If the problem is with the station wiring or telephone instrument, however, the serviceperson will not make the repair. (But of course you will be billed for the service call.)

If this is the case, what benefits are there to renting instruments from the local telephone company?

In some areas, local telephone companies still provide free home telephone repair for the elderly and handicapped provided they are renting their instruments. (You should first check with your local telephone company to find out if this is part of its service.) Renting phones also allows those who may not be able to afford buying a more expensive instrument to rent one instead at a small monthly charge.

Local telephone companies maintain phone stores to do repairs or to replace telephone instruments. While renting, you have guaranteed quick repair service if you take the instrument in to its phone store.

When buying your own phone, consider the service warranty, which can vary from 30 days to one year. Know how service and repair will be handled. Will the company you buy from repair your telephone at a locally maintained phone store or will it require you to mail it (and take a month or two to fix it)? Will the company that sold you the telephone be in business a year from now? If you plan to spend a great deal of money for a telephone, it is wise to check out these items first.

Buying your telephone eliminates the monthly rental fee that, after a few years, would add up to much more than its cost, which will save you money in the long run. But make sure you buy a telephone that will pass the breakeven cost versus the renting charges, or it might end up costing you more to buy.

CHAPTER TWO

Telephone Services Available

There are a number of telephone services available to the home telephone consumer, ranging from answering machines to Call Waiting to Three-Way Calling. This chapter provides a comprehensive guide to the selection and use of quality telephone services.

Answering Machines

Although an answering machine may be the solution to one set of problems, it may very well create a whole new set of headaches for the uninformed consumer.

The installation of certain types of telephone answering machines will immediately cause your telephone circuit to go haywire. Figure 2-1 shows two telephone numbers wired at the same phone modular connector that you intend to use for your answering machine. As with telephones, the answering machine's standard pin connections to the telephone circuit are pin positions 3-red and 4-green. However, this type of answering machine has a four-pin line cord, and pin positions 2-black and 5-yellow also will make contact with the phone modular connector pin positions 2-black and 5-yellow. With your second telephone number connected at pin positions 2-black and 5-yellow, the trouble is caused when the first telephone number is called and your machine answers the call. It will short the leads in the line cord at pin positions 2 and 5 since this type of answering machine is designed for PBX or Key telephone systems, where a button lights up to signal that a line is in use. In a home situation, when the first telephone number is called and your machine answers the call, it will automatically short the second telephone number and cross it with the first number, placing both numbers in trouble every time the machine answers the first telephone number.

There are three methods for avoiding this particular situation (Figure 2-1). The easiest method is to purchase a telephone answering machine

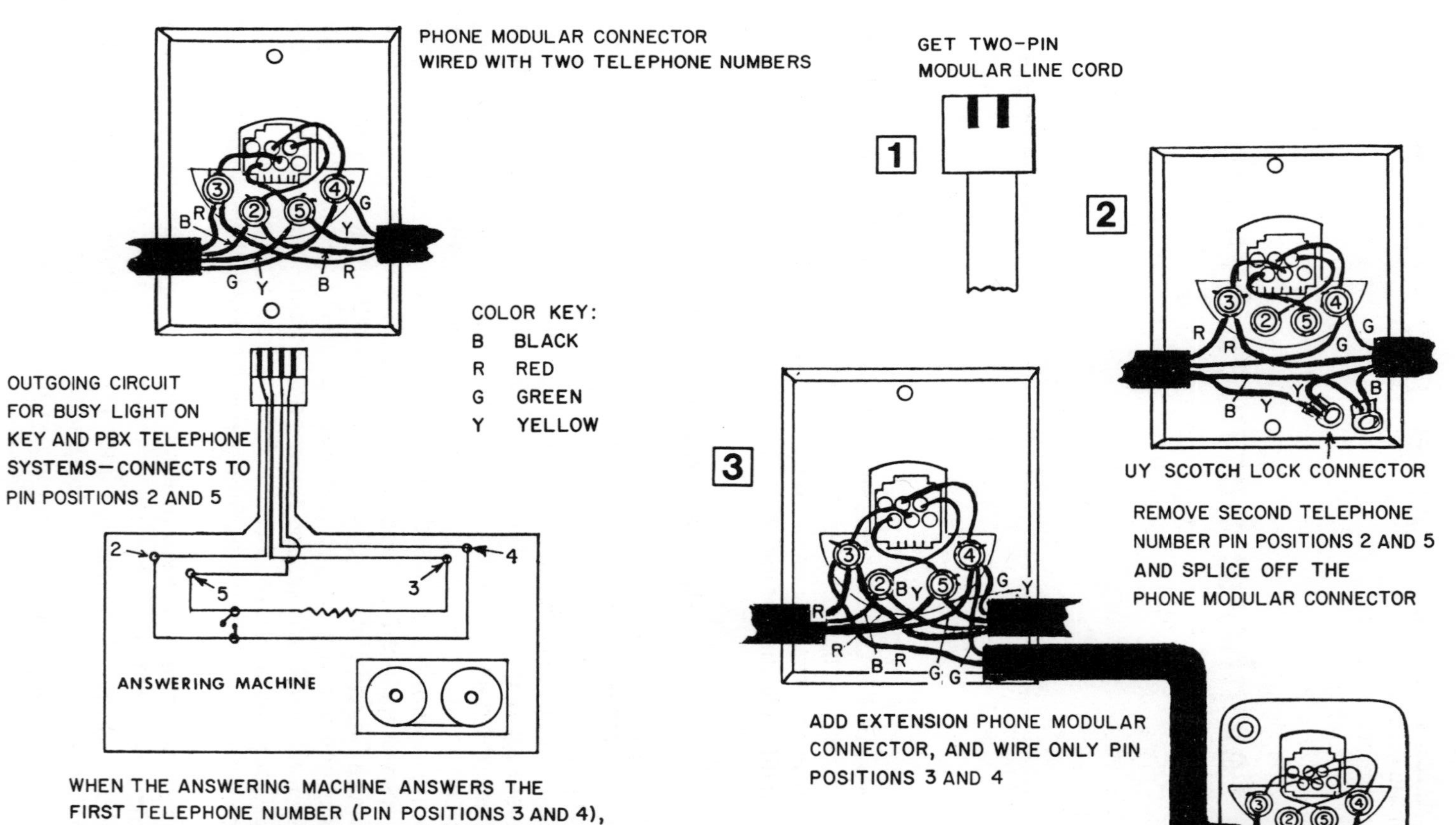

Fig. 2-1 Answering machine with 4-pin modular line cord—problem and solutions.

with only two pin positions in the line cord. (When inspecting the modular plug at the end of the line cord, you should see only two copper pins.) The second method is to remove the second telephone number pair from the phone modular connector intended for use by your answering machine and splice it back together with UY Scotch Lock connectors or a connecting block. The third method is to install another phone modular connector next to the existing one, connecting only the first telephone number to it at pin positions 3-red and 4-green. Then use this phone modular connector for your answering machine.

In today's market, a telephone answering machine can be purchased from $100 to $500, starting with a simple recorder and ending with a highly sophisticated message center. The single-tape systems are the least expensive and require prerecorded outgoing messages spaced approximately 30 seconds apart for as many incoming messages as you expect to receive. Usually, one side of the tape is prerecorded at the factory in a professional manner and the reverse side left blank so users can record their own messages. When playing back incoming messages whether at home or through remote control, you will have to first pass through all your prerecorded outgoing messages. To change the outgoing message, you will need to record the message for as many times as you expect to receive incoming calls.

The single-tape systems require a longer period of time to operate than double-tape systems and can be inconvenient if you recall your messages by remote control or change the outgoing message frequently. If you do choose a single-tape system, get one with removable tape so that if the tape breaks or the sound quality deteriorates it can be replaced. Answering machines using nonremovable tapes need to be sent to the shop for repair.

Double-tape systems are more expensive but much more convenient to operate. Two separate, removable tapes are used, one for outgoing messages (only one prerecorded message required) and one for incoming messages. A removable outgoing tape is convenient because it allows different prerecorded messages to be easily substituted at appropriate times. Also, when you play back incoming messages, there is no need to pass through time-consuming outgoing messages, and important incoming messages can be saved and stored for later use. With double-tape systems that have remote control, it is possible to change outgoing messages from outside the house.

The remote control feature on a telephone answering machine allows you to retrieve your incoming messages at any time from any telephone, which means you will be able to return calls promptly. Some remote controls require a beeper device while others can be activated by another touch tone phone. With some, you can save or erase messages, record a new one, even leave yourself a message—all via remote control. Some of the newer remote controls feature a user-codable access code that eliminates the possibility of others accessing your telephone answering machine. With the Remote Message Indicator feature, if the machine has recorded any messages since the last time you checked, it will answer on the first or second ring, but if there are no new messages, the machine won't answer until the fourth or fifth ring, allowing you time to hang up and save the cost of a call.

Two-way recording, another option, allows you to record your two-way conversation at the push of a button; this is important if you are receiving complex instructions from a caller or if a conversation may have legal implications. The two-way recording option should come with an announcement beep approximately every 14 seconds, indicating that the conversation is being recorded since recording two-way conversations is illegal without the other party's permission.

The ring control switch allows you to control the number of incoming rings needed to activate the answering machine. Some have Voice Operated Switching (VOX), through which the machine shuts itself off five or six seconds after the caller has hung up, saving dead space on message tapes that are set up to run for a certain message length. Some answering machines turn themselves off after a calling party hangs up with a mechanism referred to as Calling Party Controlled. The problem with this option is that it requires a special signal to be sent from the local telephone company, and not all companies generate this signal.

Most answering machines have some sort of indicator to let you know if the machine is on or if you have received messages. Some will even let you know how many messages have been recorded. A Ringer Equivalent Number (REN) helps determine if your instruments' combined REN exceeds the ringer equivalence of five; if it does your telephones won't ring.

It's a good idea to call yourself from time to time after your telephone answering machine has been installed to check sound quality and operation, which may deteriorate after installation.

Call Waiting

Call Waiting is a service that frees you to use your telephone and still receive incoming calls. If you are on the phone when another person calls, that caller will hear a normal ringing while you hear a beep or click, letting you know that someone is calling. To answer Call Waiting, you simply depress the hook switch, which you can do as often as you like, allowing you access to both callers. In case you don't hear the first beep, there will be a reminder beep after ten seconds. In addition, about ten seconds after you finish talking with the person whose call came in on Call Waiting and switch back to the original caller, you will hear a click. When talking to someone else who gets a beep for Call Waiting on their line, you will again hear a click.

With Call Waiting, you won't miss a call because you are talking on your telephone, and you don't have to worry about long conversations tying up your line. One disadvantage, though, is that if you choose not to answer Call Waiting, the second caller may assume you are not home because he or she hears the regular ringing sound. Also, if you receive a Call Waiting call while you are dialing out, the calling party will then get a busy signal.

The Call Waiting service usually comes in a package with Call Forwarding, Three-way Calling, and Speed Dialing. Your local telephone company needs to have special switching equipment in order to offer these services, so they are not available in all locations. If your telephone company cannot offer these services, check the upcoming section on Field Exchange Circuits.

Call Forwarding

Call Forwarding makes it possible for your calls to reach you at another telephone no matter where you are.

To activate Call Forwarding:

1. Dial 72 (plus # on touch tone telephones).
2. Listen for the dial tone.
3. Dial the forwarding number.
4. Confirm when your party answers.
5. If there is no answer, repeat the first three steps, listen for two clicks, and then hang up.

To cancel Call Forwarding, simply dial 73 (plus # on touch tone telephones). Activating (canceling) Call Forwarding must be done at the phone whose number is to be (was) forwarded.

An alternative to an answering machine, Call Forwarding can ensure that you don't miss calls just because you're away from home; you can forward your business number to your home and vice versa.

It is important to remember to cancel Call Forwarding when you are finished using it. If you forget that your number has been call forwarded, you may end up calling for telephone repair, thinking when your phone doesn't ring that something is wrong with it. Furthermore, if the line your calls are forwarded to goes into trouble, the telephone company will have to perform a few tests before it determines that the trouble is on a forwarding line. When this happens, you'll end up with a repair charge.

When a telephone number is put on Call Forwarding, the owner is charged for the call as it leaves the home. If calls are forwarded to a long-distance number, you will be charged at long-distance rates.

Three-Way Calling

With Three-Way Calling, it is possible to talk to two different people at the same time. Great for conference calls, this service allows you to share long-distance calls from family or friends with someone else in your area that both you and the caller would like to talk to. By using Three-Way Calling, you can save the long-distance caller the expense of making an additional long-distance call.

Certain steps need to be followed in order to activate Three-Way Calling:

1. Dial the first party.
2. Depress the hook switch briefly.
3. Listen for the dial tone.
4. Dial the second number; you are now connected privately to the second party.
5. Depress the hook switch again to add the first party to the second party.

To cancel Three-Way Calling:

1. The first party you call can simply hang up the telephone, leaving you in a two-way conversation with the second party.
2. To disconnect the second party, wait until he or she has hung up and then depress the hook switch briefly.
3. If you try to add a second party but there is no answer or you get a busy signal, depress the hook switch briefly twice and you are back to the original party.

If you happen to receive a call while you are using Three-Way Calling, that caller will get a busy signal.

Speed Dialing

Speed Dialing is actually a memory dialer located at your local telephone company's central office, where two ranges of memory capability are usually offered: one holds eight telephone numbers with one-digit recall codes and another holds 30 numbers. Both ranges would allow you to store up to 38 different telephone numbers.

Purchasing a memory dialer should not be necessary, though, if you already have Call Forwarding or Three-Way Calling.

To activate Speed Dialing:

1. For Speed 8, dial 74 (74# on touch tone).
2. For Speed 30, dial 75 (75# on touch tone).
3. When you hear a dial tone, dial first the one-digit code (for Speed 8) or two-digit code (for Speed 30) and then the telephone number you want to store.

To use Speed Dialing, simply dial the appropriate code and the complete number will be dialed for you. To cancel Speed Dialing, repeat step one or two and then hang up.

Rotary Hunt Group

The Rotary Hunt Group is a service that can benefit small businesses that require two to ten telephone lines to conduct business. For the Rotary Hunt Group service to work, the local telephone company at the central office connects groups of telephone numbers (generally two to ten) so that when a call is placed to one of the group's listed numbers, the call will consecutively hunt through the group to find an idle line. It is ideal for those businesses that require only two to five lines and don't want to pay the expensive installation and monthly rental charges for PBX or Key system telephone instruments.

This type of service is becoming more popular since more and more people today are operating businesses from their homes. Actually, the only disadvantage of the Rotary Hunt Group service is that you cannot transfer calls to other telephones or put incoming calls on hold. However, there are telephone instruments available that would work well with this service as they can handle multiple telephone numbers and can put calls on hold (see Chapter One, Multiline Telephones).

Field Exchange Circuits

If your local telephone network's central office lacks the necessary switching equipment to provide Call Waiting, Call Forwarding, Three-Way Calling, Speed Dialing, or touch tone calling, the installation of a Field Exchange Circuit will connect you to a nearby network office where these services are available (Figure 2-2).

Referring back to Figure I-13 in the Introduction, which shows five local telephone networks, you'll notice that all the telephone cables (except the trunk cables) originate at local central offices within their own local telephone networks. A Field Exchange Circuit, on the other hand, is a telephone circuit that originates at a central office outside of the local telephone network where it operates. It can also originate at a central office outside of the service area where it operates. (Figure 2-2 shows the routing connections for a Field Exchange Circuit.)

In the telephone business, a Field Exchange Circuit is considered special because it may require individual engineering before its installation can take place, sometimes requiring the assistance of personnel from two or more networks. In some cases, the engineering and installation can take weeks. If you are considering the installation of a Field Exchange Circuit, first refer to Chapter Five, "Special Types of Telephone Trouble and Repair—Field Exchange Circuits."

If you have a Field Exchange Circuit installed to add those services not available through your local central office, you will have to change your telephone number to one determined by the new central office. Also, what used to be local telephone calls (both incoming and outgoing) may be more expensive because your Field Exchange Circuit is not in the same local area where you live. (Verify these expenses by checking with your local telephone company's service installation department.)

An advantage to having a Field Exchange Circuit is that it often allows you to maintain the same telephone number that you may have had for years or have had printed on business cards when moving from one local telephone network area to another. However, if you move a great distance away, a Field Exchange Circuit may not be practical, considering the additional expense for what would be local calls in your new area.

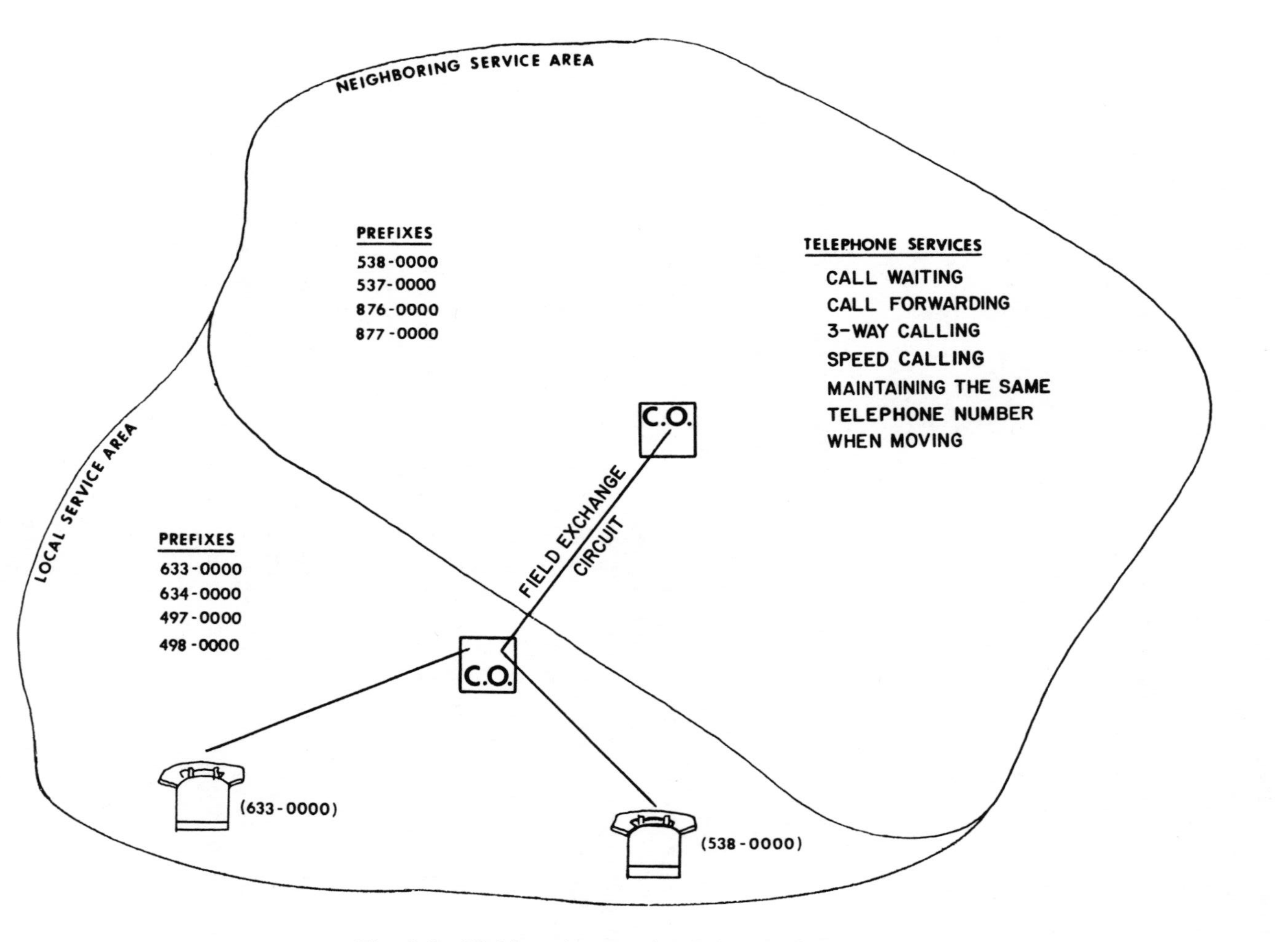

Fig. 2-2 Field exchange circuit for services.

If most of your incoming and outgoing calls are to and from a local telephone network area outside your own, all your calls would be like local calls in the area where your Field Exchange Circuit originates (Figure 2-3). Some telephone companies provide special calling charges for a certain number of outgoing calls placed to a particular area, so it would be wise to compare the cost to your calling needs. Since a Field Exchange Circuit allows incoming calls from the area in which it originates to also be local calls, this is one advantage to the people who call you from that area because their calls will be local and not long-distance.

Off-Premise Extension

An Off-Premise Extension makes it possible to have the same telephone number installed at two different locations (Figure 2-4). If the primary location is your home while the secondary location is your downtown business office, this service would allow all calls to your home to ring at your business also; conversely, if your business is the primary location and your home secondary, all calls to your business would then ring at home. With the home as the primary location, Off-Premise Extensions are commonly connected to workshops and boats when docked.

An excellent alternative to Call Forwarding, the Off-Premise Extension does not require you to be at the primary location to program the telephone number required for the secondary location, as Call Forwarding does, and is usually less expensive than installing the additional number necessary with Call Forwarding. Off-Premise Extensions are a practical way to go if Call Forwarding is not available in your local telephone network area.

If the secondary location for the Off-Premise Extension is outside your local telephone network area, the circuit will probably require special engineering since it will need to be connected much like the Field Exchange Circuit, which was described earlier. If you're considering the installation of an Off-Premise Extension, first refer to Chapter Five, "Special Types of Telephone Trouble and Repair—Off-Premise Extensions."

To connect this type of service, a one-time installation charge for the secondary location and an additional monthly service charge for maintaining the circuit from your local telephone company will be required.

Permanent Answering Service

To add a Permanent Answering Service to your home telephone system requires the telephone company to install a circuit from its central office to the private answering service switchboard. This type of installation is similar to the Off-Premise Extension, but instead of a telephone instrument at the secondary location there is a switchboard capable of receiving hundreds of telephone circuits. The answering service hears your phone every time it rings at home and will answer on the third or fourth ring or as instructed.

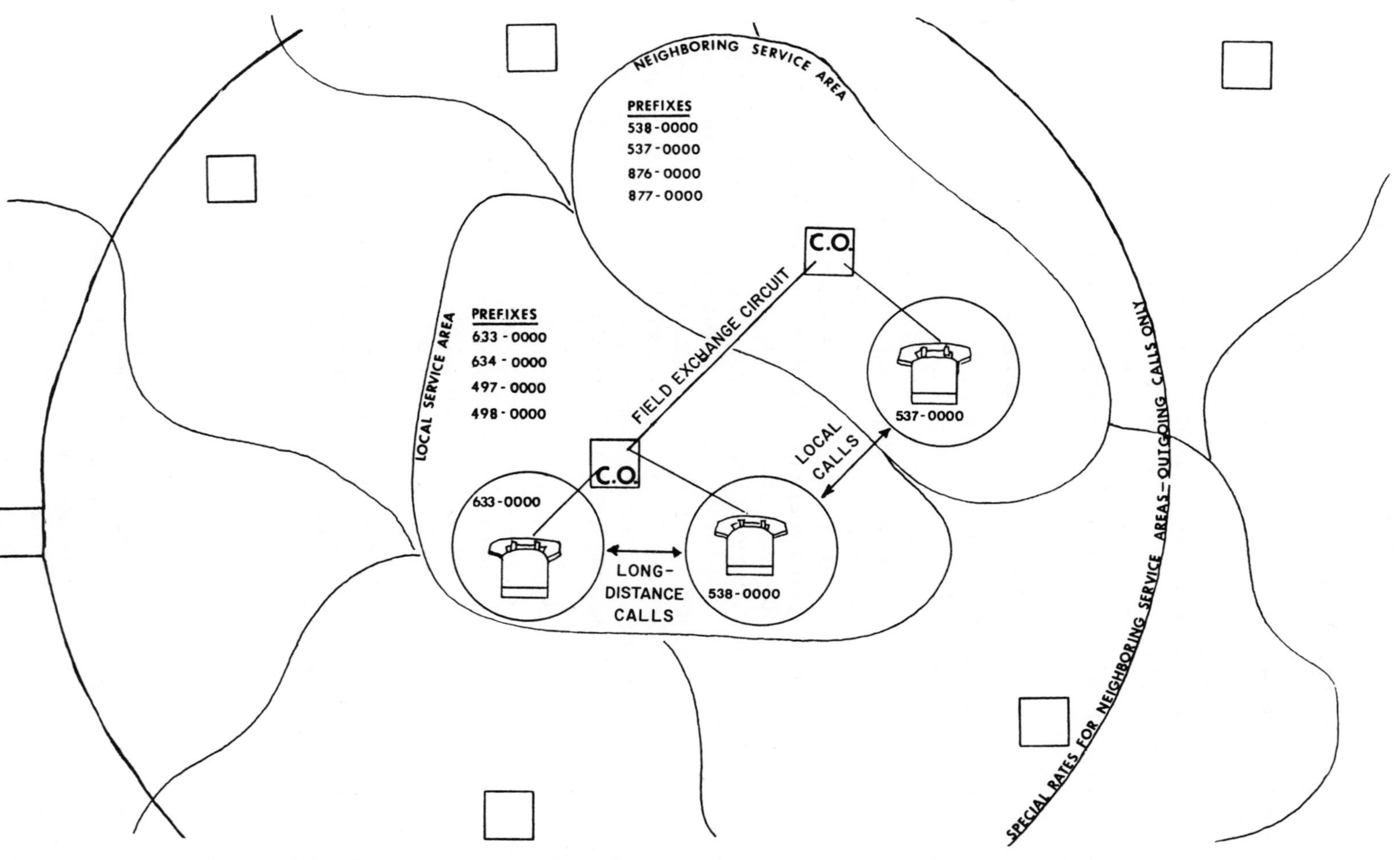

Fig. 2-3 Field exchange circuit for local calling to and from a neighboring service area.

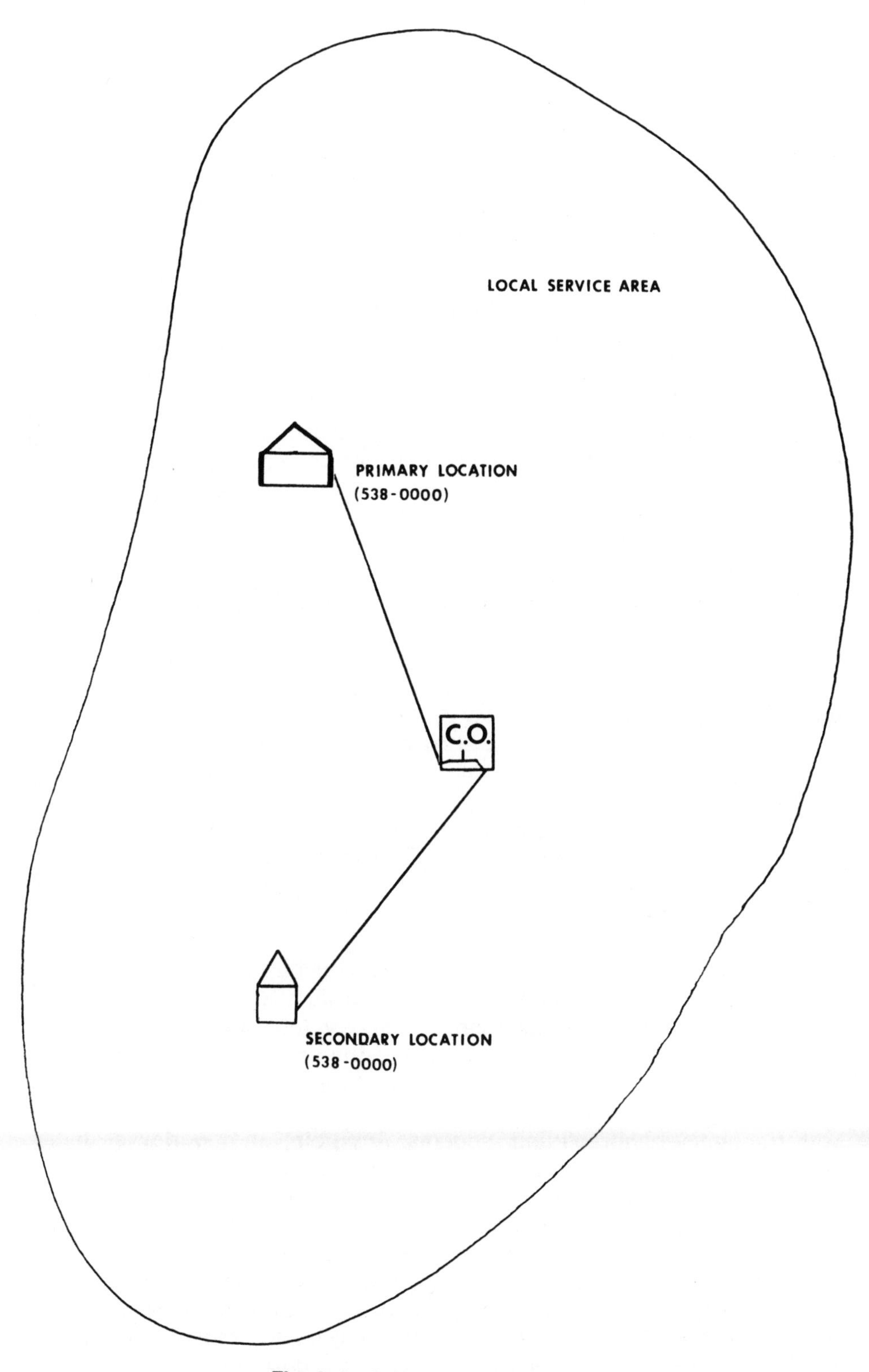

Fig. 2-4 Off-premise extension.

There will be a one-time installation charge for the connection to the answering service switchboard and an additional monthly service charge from your local telephone company for maintaining the circuit. In addition, there will be service charges from the private answering company for taking and relaying your messages.

If you're considering the installation of a Permanent Answering Service, first refer to Chapter Five, "Special Types of Telephone Trouble and Procedures for Repairing—Permanent Answering Services."

Data Circuits

Currently, the only method for linking your home computer terminal with other computers is by means of your telephone cable pairs, which function as one type of data circuit.

For one computer terminal to send or receive information to or from another, the information is sent by digital code, which cannot be transmitted over telephone cable pairs. Therefore, a modem is required to translate digital codes to a modulated carrier waveform that can then be transmitted over telephone cable pairs. (Figure 2-5 shows a simple diagram of the modem's translating characteristics, known as modulating and demodulating.) Selecting the proper modem for your needs depends on the type of data circuit and computer service you wish to connect with.

The major points of concern in the data communications field for transmitting data over telephone cable pairs are the speed at which the data can be sent and the accuracy of the data received.

There are three basic types of data circuit links available with telephone cable pairs: simplex, half-duplex, and full-duplex, illustrated in Figure 2-6.

With simplex, transmission is from A to B only and not from B to A. With half-duplex, transmission is possible from A to B and then from B to A but not simultaneously. If a two-wire circuit is used, the line must be turned around to reverse the direction of transmission. With full-duplex, transmission is possible from A to B and from B to A simultaneously, eliminating the turnaround time required with the half-duplex. Also, a two-wire circuit can be engineered to provide full-duplex transmissions.

There are two types of data circuitry access: dial-up lines and leased lines. The dial-up lines, two-wire circuits connected by the local telephone network and switched through the central office, are used for half-duplex transmissions, although they can accommodate transmitting full-duplex with frequency band splitting modems. Any point on a worldwide telephone network is accessible on dial-up lines, and costs are limited to the time the lines are actually used. Because the lines are switched through telephone company central offices, they may be noisy, and computers or terminals can lose or misinterpret data due to noise. If local telephone network service is poor (e.g., static, cross-talk, disconnects), it will equally affect your data communications. Since this type of data link is switched through central offices, data transfer rates may be affected by turnaround time and connect and disconnect times. Also, delay and distortion are

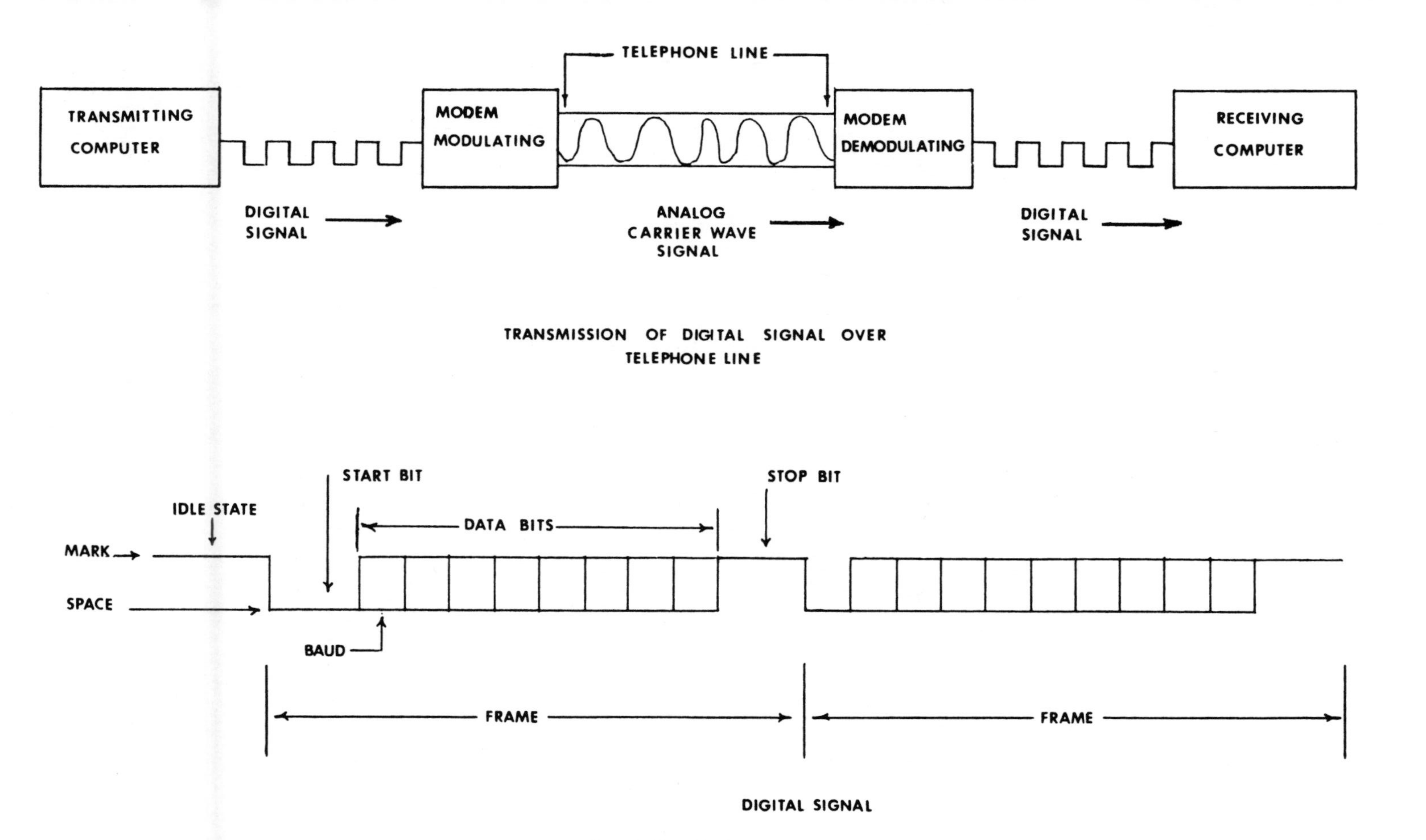
TELEPHONE LINE
TRANSMITTING COMPUTER
MODEM MODULATING
MODEM DEMODULATING
RECEIVING COMPUTER
DIGITAL SIGNAL
ANALOG CARRIER WAVE SIGNAL
DIGITAL SIGNAL
TRANSMISSION OF DIGITAL SIGNAL OVER TELEPHONE LINE
START BIT
STOP BIT
IDLE STATE
DATA BITS
MARK
SPACE
BAUD
FRAME
FRAME
DIGITAL SIGNAL

Fig. 2-5 Modem.

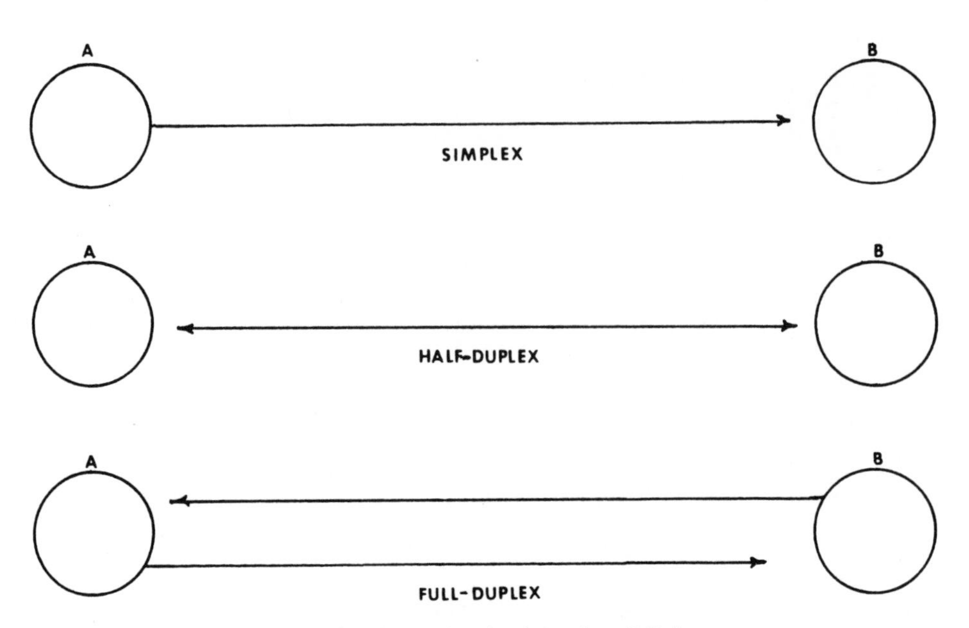

Fig. 2-6 Three basic data circuit links.

caused by data signals that are transmitted at nonuniform speeds due to different distances and slight frequency differences between one central office and another.

Although more costly than dial-up lines, leased lines greatly reduce the problems associated with dial-up lines because they are directly connected at both ends of the circuit and require no switching. Ready availability, freedom from busy signals, precise engineering for better data quality, and faster transmission rates are the highlights of leased lines, which are commonly four-wire circuits capable of half-duplex or full-duplex transmissions. Despite the disadvantages of higher cost and the fact that the line is connected to only one location, if the data communication demands of the user entail high-volume, high-quality, and high-speed traffic between two points, a leased line is the best choice.

Because a modem has much to do with the quality of data communications, a voice grade modem that is designed for use on telephone cable pairs should be selected for its type of service (dial-up or leased), data rate, and acceptable level of error performance. The two broad categories of voice grade modems are:

1. Asynchronous modems operating at a maximum data rate of 1,800 bps over dial-up data circuits and 2,000 bps on engineered leased lines.
2. Synchronous modems operating at a maximum data rate of 4,800 bps over dial-up data circuits and 9,600 bps on engineered leased lines.

Whether your data communication needs are simple or sophisticated, most telephone companies maintain a data communications department to help you determine what type of data circuits and modems you need.

Alarm Circuits

One of the more popular types of home security is the off-premise monitoring security system, which has a control unit located at your home with various sensor devices providing detection of intruders at points of possible entry. Connected to the off-premise monitoring station by your telephone cable pair, the control unit at your home is known as an alarm circuit.

To connect this type of security service, you will need to install a special type of equipment jack (called an RJ-31X) at your home along with a cable pair from your local central office to the monitoring station switchboard. (Figure 2-7 details the alarm circuit.)

Installed next to the control unit, the RJ-31X is the interface between your telephone cable pair and the control unit. It is placed on the circuit before any other telephone instruments or equipment and connects the control unit to the alarm circuit in a series rather than a branch connection. Your telephone cable pair circuit will come from the house protector or protected service terminal to the RJ-31X then go through the control unit and back to the house protector or protected service terminal, finally connecting to your inside wire.

When considering the installation of an off-premise monitoring security system, keep in mind that your local telephone company will be installing and maintaining a great part of the alarm circuit through its telephone cables. If the quality of its service is poor, the quality of the security system will be affected; this means you can expect a high rate of false alarms and moments when the alarm circuit will be out of service.

Lifeline

Lifeline, from Lifeline Systems, Inc., is a personal emergency response system that hospitals and service organizations offer to their local communities. It is especially useful for elderly and handicapped people who live alone; it provides assurance to users and their families that help is available. This program is available nationally and in Canada at more than 1,400 installations.

Lifeline consists of three parts:

1. Electronic equipment located in your home. A portable help button and home unit linked to your telephone are used to summon emergency assistance.
2. An emergency response center, located at the hospital, where trained personnel provide 24-hour-a-day coverage.
3. People who have agreed to respond to your call for help. They are sent immediately in the event of an emergency.

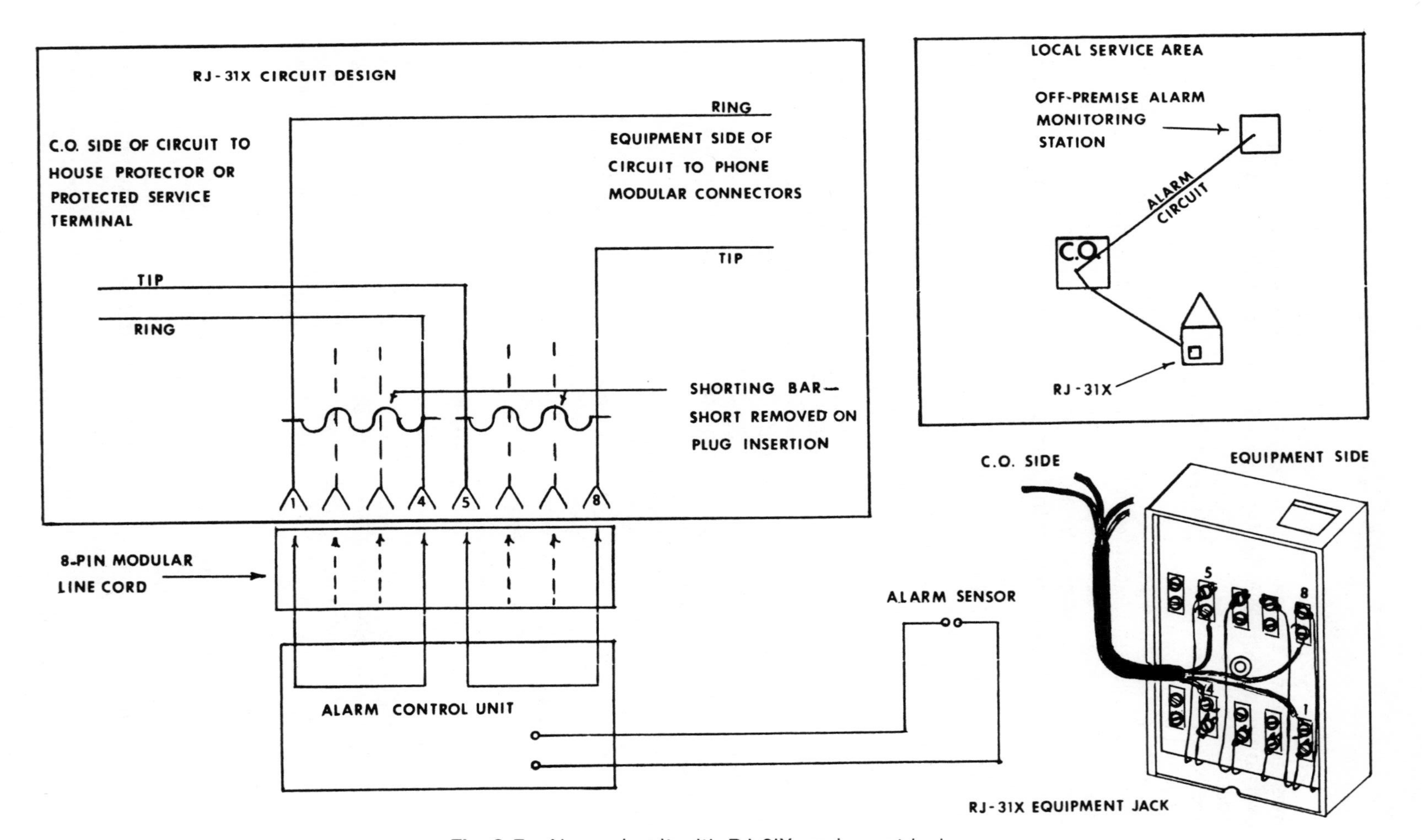

Fig. 2-7 Alarm circuit with RJ-3IX equipment jack.

The Lifeline system begins when you push a small help button that you wear on a chain around your neck or on a strap around your wrist. Whenever you need help, day or night, you just push the button to contact the hospital's emergency response center. Your call for help activates your home unit attached to your telephone, which automatically dials the hospital. The Lifeline home unit can work even if your phone is off the hook or power is out. When coordinators at the emergency response center receive your signal for help, they will immediately try to reach you by telephone to see what help you need. If they are unable to reach you, they will send a responder to your home. You have selected the responder in advance—a friend, neighbor, relative, or anyone else you choose. Upon arriving at your home, the responder signals the hospital by resetting the Lifeline unit. The hospital will then call the responder on your telephone to see what kind of help you need. If you need medical help, the responder can either take you to the hospital or have an ambulance sent to your home.

Lifeline has another important feature. It will call the hospital for help even if you cannot use the help button. The home unit includes a timer, which is reset automatically every time you use your telephone. If you become unconscious or are otherwise unable to reset the timer, Lifeline automatically calls the hospital and starts the emergency response procedure for you.

For areas that have no emergency response centers, Lifeline Systems, Inc., intends to provide Lifeline Central, a centrally located response center. For details about the Lifeline system, contact Lifeline Systems, Inc., 1 Arsenal Marketplace, Watertown, Massachusetts 02172 (1-800-451-0525).

Sensaphone 1000

The Sensaphone 1000, from the Gulf & Western Company, monitors your home or office, even when no one is there. It automatically monitors your smoke detector and other security devices for fire, temperature changes, and power failures. If an emergency occurs, Sensaphone automatically calls four of your preprogrammed numbers in sequence until answered and explains to the listener the nature of the problem. Action can then be taken within minutes. If you are home, Sensaphone issues the same warning through its speaker.

The Sensaphone 1000 telephone instrument can be found in many discount stores. For more information, contact the Gulf & Western Company.

BSR X-10

The BSR X-10 unit offers remote control of home appliances. You can turn lights, ovens, sprinklers, radios, and other appliances on and off without being home. Connect the unit's switching modules to the appliances you wish to control remotely; then simply call home, hold the remote unit up to the receiver, and press the appropriate buttons.

The BSR X-10 unit can usually be found at department stores.

SP200 Series Software and Data Protection Devices

If you transfer data from your computer to others, you may be interested in encryption devices that will protect your data from being stolen. The SP200 Series Devices, from Voyager Development, Inc., are designed to protect proprietary software and data from being illegally copied, used, or sold.

The SP200 Series Devices are small, microprocessor-based "black boxes" with up to 64K bytes of self-contained read-only memory. The software being protected requests a unique I.D. code from the software protection device. If the correct code is not returned or the device is not available, the program can freeze, abort, self-destruct, or do anything else the software author wishes. The device allows the end user of your software to make legitimate backup copies, thus eliminating the need to supply multiple disks, and reducing time, materials, and cost.

For more detail about the SP200 Series Software and Protection Devices, contact Voyager Development, Inc., 412 S. Lyon Street, Santa Ana, California 92701. Outside California (1-800-641-3000), inside California (1-800-922-3000).

Tap Detector

The Tap Detector, installed between your telephone and wall jack with standard modular connectors, detects tiny changes in line impedance. An audible signal and red warning light alert both you and your caller to outside taps, eavesdropping phone operators, and lifted extensions. A green light advises you when the line is clear. The Tap Detector doesn't interfere with normal phone operation.

For more details, contact The Sharper Image Catalog, P.O. Box 26823, San Francisco, California 94126-6823 (1-800-344-4444).

Ring-Down Circuits

The Ring-Down Circuit, a telephone circuit where two telephone instruments are directly connected to each other by cable pairs, does not require routing into the local central office and can be installed anywhere telephone cable pairs can be placed.

The emergency telephone in elevators is one of the more common applications of a Ring-Down Circuit. To activate the circuit, simply lift the handset from the switch hook; the telephone on the other end will ring until it is answered. A Ring-Down Circuit can be installed between any two points that the local telephone network reaches, such as a store and the owner's home or two stores.

If there is a lot of calling between two points, the Ring-Down Circuit is an excellent alternative to installing a regular telephone circuit because it is less expensive and provides direct access to the other end.

Line-Splitting Circuits

When a local telephone network area requires more telephone circuits than it has telephone cable pairs to accommodate, the local telephone company will usually turn to Line-Splitting Circuits.

When line-splitting equipment is placed on a telephone cable pair, one pair is made to work as two; thus two circuits are created out of one. This is accomplished by modulating and demodulating the second telephone circuit (the carrier side of the circuit) over the first telephone circuit (the physical side of the circuit), as shown in Figure 2-8. On the physical side of the circuit, a filter is placed to screen out signals from the carrier side; and on the carrier side, a modulating-demodulating piece of electronic equipment is placed so the carrier side can operate independent of the physical side.

Typically, telephone companies install line-splitting equipment (an excellent alternative to a second-party line) on your first circuit when you order a second circuit to be installed. However, it is not uncommon for telephone companies to place line-splitting equipment on your circuit to provide a telephone circuit for a neighbor; thus, it is possible that your regular telephone circuit could become the carrier side of a Line-Splitting Circuit without you ever knowing.

One problem associated with the carrier side of a Line-Splitting Circuit concerns the transfer of data communications. Since the carrier side is already modulating a carrier wave, an additional carrier wave that is sent through will sometimes arrive in such poor quality that the data will be unreadable. Also, in some situations, answering machines will not activate when connected to the carrier side of Line-Splitting Circuits.

Whining sounds, echoes, and slow dial tone response are signs that your telephone circuit has been placed on the carrier side. If you know your second telephone circuit is going to be put on a Line-Splitting Circuit, keep your answering machine on the physical side along with computer terminals for data transmissions.

Be aware that a voice grade telephone circuit you order from the telephone company may be placed on Line-Splitting Circuits that are not compatible with data transfer techniques or answering machines. Because the telephone company is concerned only that the circuit meet voice grade specifications, the consumer must be aware of the problem and the possible need to order a special circuit for answering machine or data communications needs.

Second-Party Lines

There was a time when up to five parties shared a line. Today, telephone companies are able to provide just about everyone with a private line; the Second-Party Line is fast becoming a thing of the past.

In some local telephone network areas, though, you may have to wait for the installation of an additional telephone number because facilities are nonexistent. If there are no cable pairs available, you may have to wait

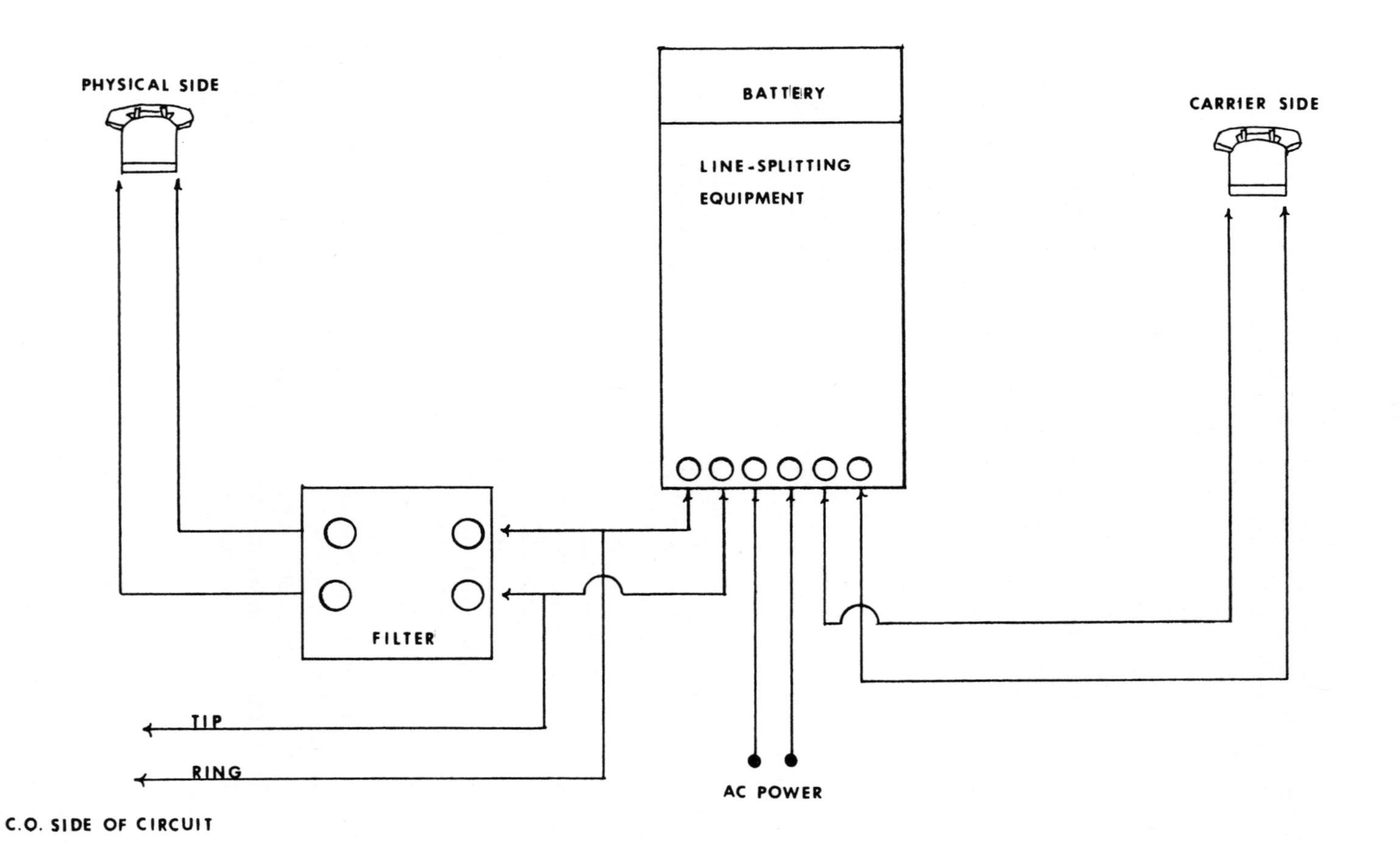

Fig. 2-8 Line-splitting circuit.

until someone in the neighborhood moves or until the local telephone company adds additional cable, which could take years. If you decide that line-splitting equipment is not suitable for you, yet you don't want to wait years, a Second-Party Line telephone number at your home may be the answer. Since Second-Party Line numbers do not require an additional cable pair, the user shares the same cable pair with the other telephone number. This other number, of course, will be the telephone number you currently have; therefore, the inconvenience associated with Second-Party Line numbers will be minimal since you control both numbers.

Because the Second-Party Line shares the same cable pair as your other telephone number, there is no need to install inside wire or phone modular connectors; thus, monthly service charges for Second-Party Lines are less than for private lines.

Long-Distance Discount Dialing Services

Since the deregulation of AT & T, a number of long-distance discount dialing services have entered the long-distance network. Because the industry is still in the process of changing to meet deregulation mandates, there will be major changes happening soon under the equal access requirements that will affect the operating methods for accessing long-distance discount dialing services.

In the meantime, long-distance discount dialing services are accessed in the following manner:

1. First you dial an access code of seven digits that routes your call to the local telephone network central office. From there, it is routed to the long-distance discount service's central office. (*Note:* Check that the service's central office is a local call and not a long-distance call since your access code could be routed to a faraway central office, resulting in long-distance charges for what should have been local calls.)
2. Once your call reaches the service's central office, you will hear a second dial tone that indicates you have reached the central office. Their switching equipment is now ready to receive your authorization code.
3. Dial your authorization code, typically five digits. This identifies you as a service user and logs the call for billing purposes.
4. Finally, dial the long-distance number you wish to reach. Your call will be routed through the service's long-distance network to another central office within the area you are calling. From here, it is routed to a local telephone network central office and at last to the person you are calling.

In order to use a long-distance discount dialing service, you will need to dial at least 22 digits. These services also require that you dial with touch tone signals unless you have one of their auto dialers that will work with rotary dialing. Accessing a long-distance discount dialing service requires more work and a good memory for long codes. Figure 2-9 details

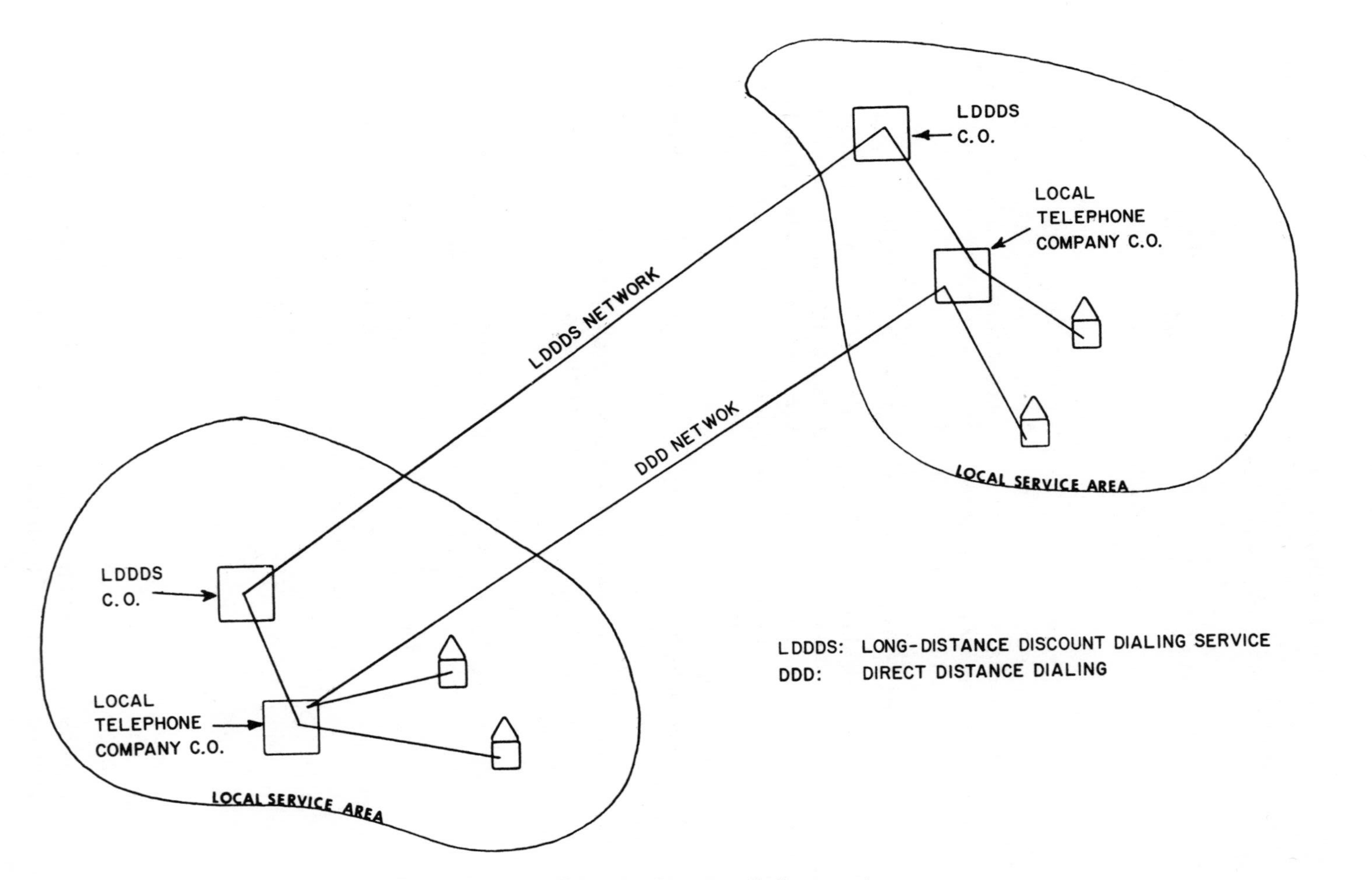

Fig. 2-9 Long-distance discount dialing services.

the network routing for a long-distance discount dialing service versus Direct Distance Dialing.

To keep your dialing digits to a minimum (ten) and not overtax your memory, most long-distance discount dialing services offer an auto dialer than will automatically dial your access and authorization codes. Because an auto dialer is connected to your telephone circuit with an RJ-31X equipment jack (see Figure 2-7), all calls, incoming and outgoing, will be routed through it. The auto dialer collects and stores all or part of the digits and then proceeds with dialing out. If you are making a local call, it will not dial the access and authorization codes and will route your call through the local telephone network, usually signaling you with two beeps. On long-distance calls, it will dial the access and authorization codes and route your call over the long-distance discount dialing service network, signaling you with only one beep.

When your auto dialer routes your call over its network and the long-distance discount dialing service's central office is busy (common during peak calling hours), it may be programmed to redial your number for up to 15 times and then reroute your call to Direct Distance Dialing (DDD, currently AT & T's long-distance network). Since 15 redials take approximately 10 minutes to complete, some callers won't wait and will try to place the call again. To confuse the user, there are no signals to indicate the auto dialer is redialing, so the user hears only dead space and may think something is wrong. Some services may program the redial feature for fifteen attempts to help ensure that you use the service and do not dial DDD. Programming the redial feature for two or three attempts would be more realistic, and this is possible with most auto dialers.

Many auto dialers do, however, have a release code informing the auto dialer that you wish to place the call through DDD, which, of course, defeats the purpose of the long-distance discount services feature.

If all your long-distance calls are to nearby area codes and are more expensive with the service than with DDD, you obviously don't need this service; however, if you are going to place calls to area codes for which the service will save you money as well as calls to nearby area codes for which the service will not, it would be wise to have the nearby area codes screened out of the auto dialer. Screening out area codes, a feature that can be programmed into an auto dialer, should be carefully checked before the installation is completed.

Because the dialing pattern of an auto dialer requires the collection of all or part of the number before it is dialed out, you will not be able to use the Speed Dialing or Three-Way Calling services provided by your local telephone company's central office in conjunction with the long-distance discount dialing service. Also, if the auto dialer hears too much circuit noise, it may interpret the noise as an on-hook signal (meaning the caller has hung up) and disconnect the line. When you answer a Call Waiting signal by depressing the switch hook to answer, the auto dialer may interpret this as an on-hook signal and disconnect the first caller. Static or switching equipment noise may also cause the auto dialer to disconnect the line.

Once an auto dialer is installed on your telephone circuit, there is about a 40 percent chance that it will malfunction within the first week and a 20 percent chance of malfunction in the second. Once this happens, your telephone circuit will be thrown into trouble. To avoid this, make sure the installer gives you instructions for disconnecting the auto dialer from your telephone circuit. Often, simply unplugging the auto dialer line cord from the RJ-31X phone equipment jack will disconnect the auto dialer from the telephone circuit and free your circuit from auto dialer–related problems.

As mentioned earlier, major changes are on the way that will affect the current method of accessing long-distance discount dialing services. These changes will allow "equal access" at your local telephone network central office, meaning that long-distance discount dialing services will be directly connected to your local telephone network, just as AT & T is today.

Once these changes have been made, you will be able to choose one company for long-distance calls by simply dialing the area code and telephone number, as you currently do with AT & T. In addition to your first choice, you may also arrange with another long-distance dialing company to use its service on a call-by-call basis using Company Code Dialing, a new telephone terminology because DDD will no longer refer exclusively to AT & T. With Company Code Dialing, you dial a five-digit code for the long-distance service you wish to use followed by the long-distance area code and telephone number.

Until these changes take place, it is impossible to know if quality of the new services will be sacrificed to quantity. Some questions that should be considered include: Will these services offer operator assistance for immediate credit for wrong numbers dialed or poor connections? Will these services compete with the quality of service provided by AT & T, a company that pioneered the long-distance network? Will these services be able to provide credit card dialing?

Cellular Mobile Telephone Service

The cellular mobile telephone service is the latest advancement in automobile telephone communications, a major improvement over earlier mobile communications where the operating service area was about a 60-mile radius from a high-powered transmitter (Figure 2-10).

The mobile telephone network operates similarly to CB radio, where a large number of users share the same open channel. In order to place a call from your car, you need to wait for the channel to clear, then call a mobile telephone operator (via a microphone), who will then dial the number for you. Once the call is answered, your conversation will be transmitted for all who are on the same channel to hear.

The cellular mobile telephone service, however, operates in a much more sophisticated network that covers approximately the same area as the mobile telephone. Instead of one high-powered transmitter, however, the service area is divided into hexagonal cells, each with its own transmit-

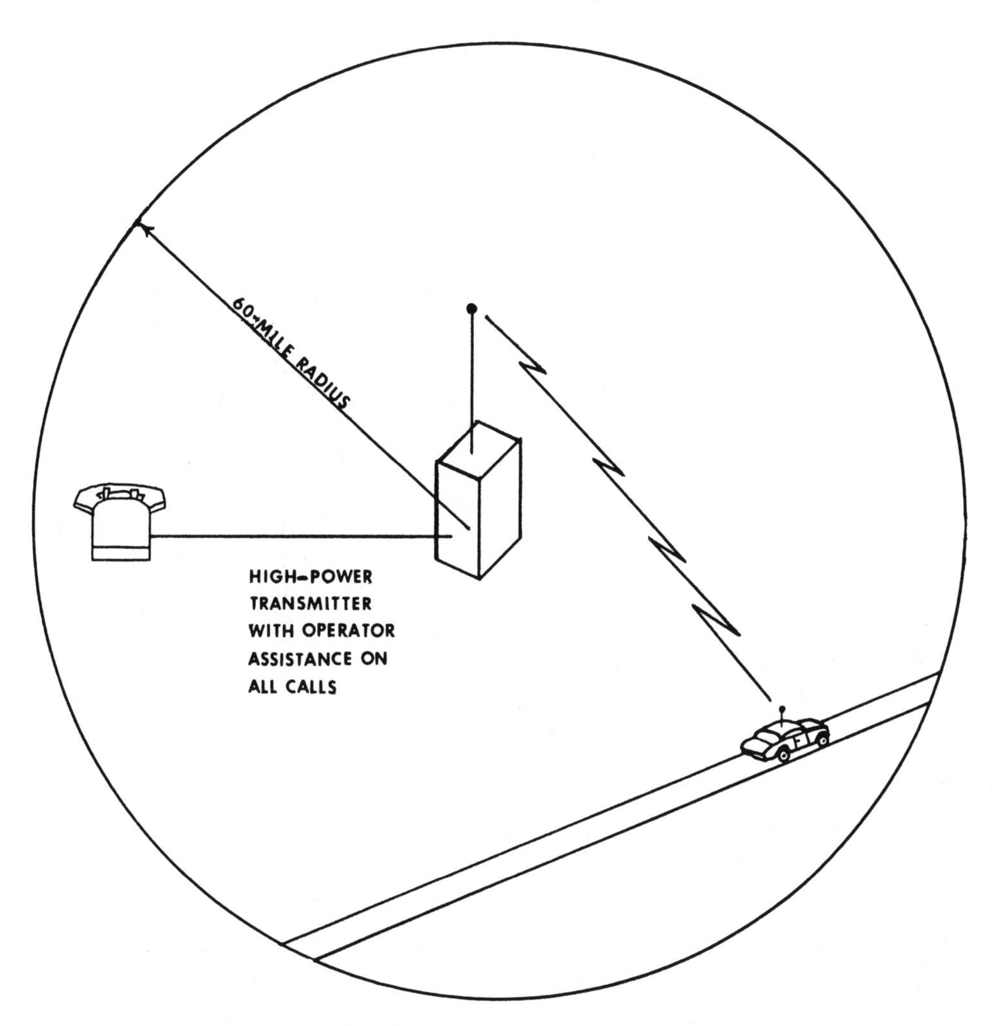

Fig. 2-10 Mobile telephone service area network.

ter and computer controller. The network of low-powered transmitters is connected by telephone cable pairs and controlled by a central office that permits many more calls than the mobile telephone network (Figure 2-11).

As an automobile travels from one cell to another, computers automatically switch the signal to the next transmitter and continue doing so as the automobile moves along through the network, all without any interruption in conversation. Thus, automobile A can travel from location 1 to 5 while the conversation continues. Each cell can handle hundreds of conversations at any one time so that automobile B traveling into the same cell as automobile A in no way interferes with it.

Cellular telephone networks today are operating in metropolitan areas, and the consumer may choose between the local telephone company or a private outfit.

At present, the different cellular mobile telephone service area net-

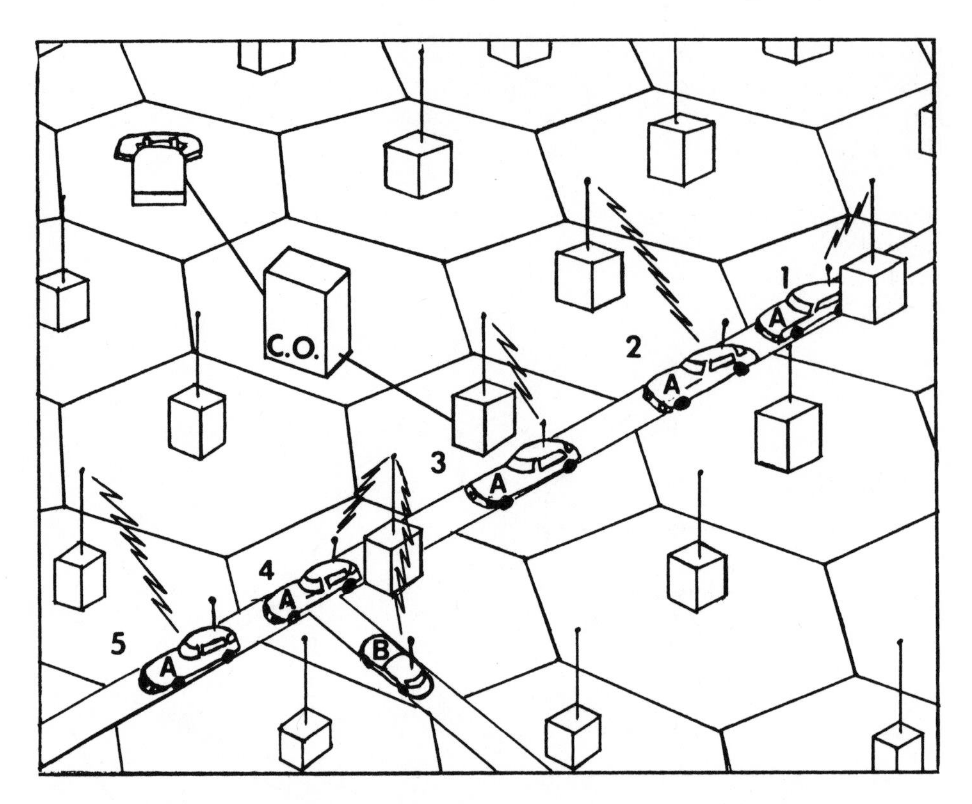

Fig. 2-11. Cellular mobile telephone system network—60-mile area—showing many cells with low-power transmitters.

works are not commonly connected, so traveling from one service area to another requires planning. It is necessary to call your cellular company to let it know what areas you intend to travel to, along with the dates. It, in turn, will provide you with temporary dialing codes for those service areas so it can recognize your call. (Designs to connect different service areas together are on the drawing board, so the planning and special codes necessary today may eventually become obsolete.)

Cellular telephone instruments operate just like home phones. A telephone number is assigned to your cellular mobile telephone, and you can be called or place calls just as you would at home. Also, many telephone services may be added to your cellular telephone system, such as Call Forwarding, Three-Way Calling, Speed Dialing, and Call Waiting; it is also possible to access the long-distance discount dialing services. There is even a hand-held cellular telephone instrument available with all the same functions for people who require even more mobility.

The cost of the cellular mobile telephone instrument runs from a low of $1,600 to a high of $3,100, depending on the features you require. Two features I recommend for driving safety are memory dialing (you spend less time dialing and more time watching the road) and the speaker phone (you can carry on a conversation with both hands still on the steering wheel). Billing for a cellular mobile telephone network generally runs about $45 a month, in addition to 27 cents to 45 cents a minute, depending on the time of day. The per-minute charge occurs both for calls placed and for calls received.

With more and more cellular mobile telephone instruments being installed, there may eventually be a rash of robberies, as there were with auto tape decks. At the current prices, it might be worth the extra few dollars a month to add your cellular mobile phone to your auto insurance policy.

CHAPTER THREE

Telephone Installations for Single-Unit Dwellings

This chapter presents all the information you need to install a telephone in a single-unit dwelling. Each connecting point—from the unprotected service terminal to the phone jack—is covered. In addition, advice on how to avoid unnecessary inconvenience and expense is offered.

There are five connecting points to check when installing a telephone in a single-unit dwelling: the unprotected service terminal, drop wire, house protector, inside wire, and phone jack (see Figure I-7 in the Introduction).

In most cases, when your moving into a new home, all these connecting points will already exist, having been installed by a previous owner, tenant, or building contractor; however, they might not be in the condition or arrangement necessary for your purposes. By understanding each connection, you will more easily be able to determine what may be needed in terms of repairs, changes, or additions.

Unprotected Service Terminal—Aerial Type

Locating your aerial service terminal is done by following the drop wire from the house back toward the telephone cable (see Figure I-8 in the Introduction). Generally, the closest terminal to the house will be yours.

The installation of an aerial service terminal or aerial drop wire must be left to the telephone company. Because the installer may need to reach the service terminal (e.g., if drop wire must be placed), you should make sure it is accessible. It may be necessary to alert your neighbor that the installer needs to get into his or her backyard in order to reach the service terminal; otherwise, locked gates and unfriendly dogs may lead to needless rescheduling hassles.

Figure 3-1 shows the required climbing space for an installer to reach the service terminal on the telephone pole. If trees and vines obstruct climbing space so that a ladder cannot be safely placed at the service

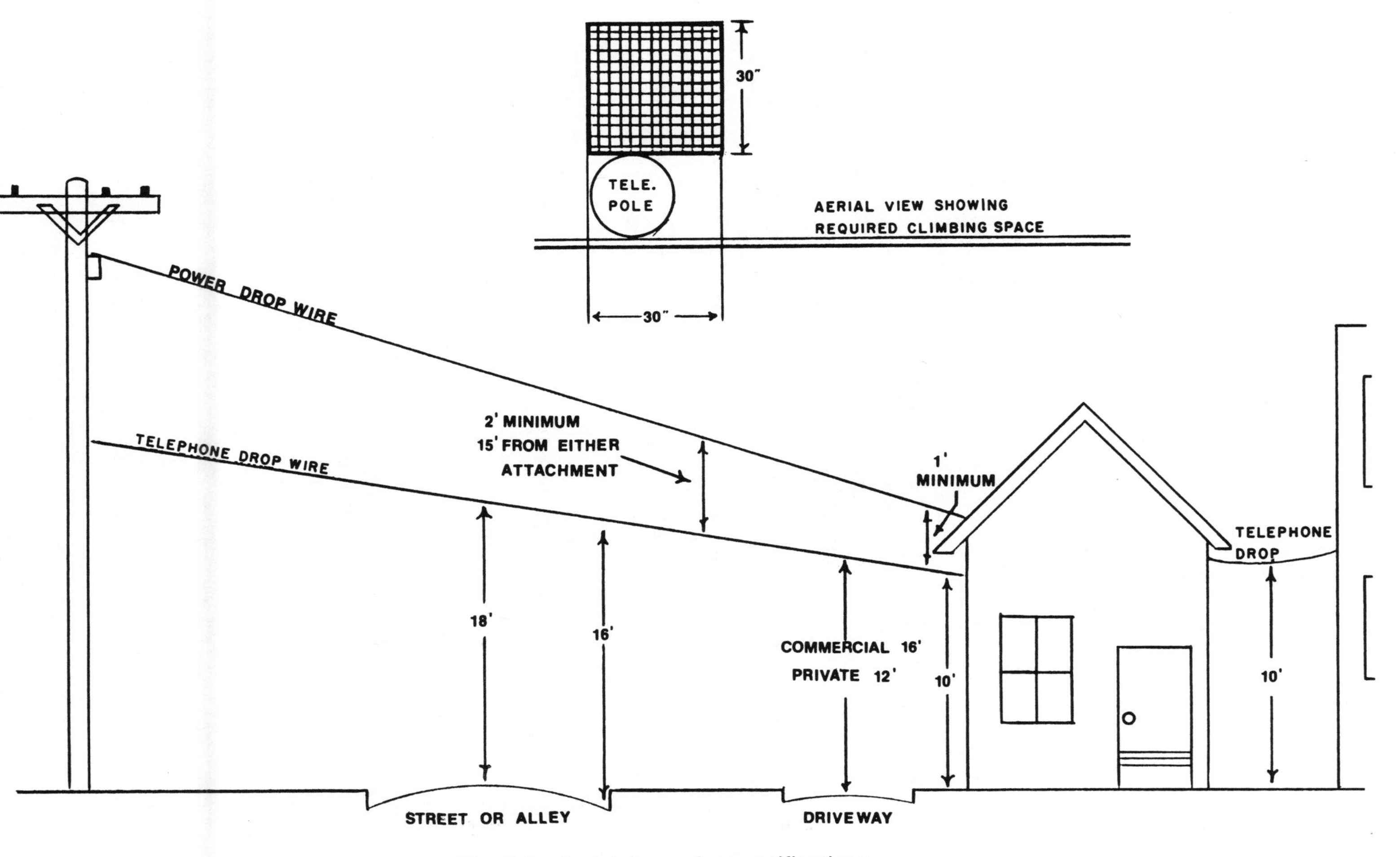

Fig. 3-1 Aerial drop wire specifications.

terminal, clearing will be necessary. The responsibility of keeping the climbing space clear belongs to the owner of the yard where the pole is located. Because neighbors are not always pleased about the added expense of clearing away obstructions for someone else's phone, leave it to the telephone company to contact your neighbor to set up an appointment for clearing.

While checking access requirements, note the service terminal's general condition. A missing cover, hanging wires, or a bird's nest create problems and should be taken care of immediately.

Unprotected Service Terminal—Buried Type

The buried service terminal, found in areas where telephone cables are also buried, are of three basic types: the permanently buried, hand box, and pedestal.

The permanently buried service terminal usually has two or three buried drop wires connected to it for service to your home, which means there will be enough pairs to accommodate anywhere from six to twelve telephone circuits.

One reason for installing up to three (three-pair) buried drop wires along with the service terminal is to try to avoid the time and expense of digging up the terminal later. Adding or repairing buried drop wire would require digging up the service terminal. (Having one or two alternative drop wires already installed reduces the chances that you'll have to repair one.) In addition, when installing additional telephone circuits, you won't need to worry about buried drop wire installation.

If it does become necessary to dig up the terminal, have the local telephone company locate it for you so that you won't accidentally damage it while trying to find it. (Damage expenses can be substantial.)

Because permanently buried service terminals and telephone cables are out of sight, they do not distract from the beauty of a neighborhood. The obvious disadvantage, though, is that if they should become damaged, the area will have to be dug up in order to repair them.

The hand box buried service terminal (Figure 3-2) is accessible simply by lifting a cover plate, allowing telephone personnel to change the main or feeder telephone cable pairs and test serviceability with relative ease. Commonly placed out of sight at the front of the house, near the street, this type of terminal eliminates the need to do any digging for changes or repairs, but if it is not made airtight after accessing, the services terminal and hand box may fill with water, affecting telephone service.

Most commonly used, the pedestal buried service terminal (Figure 3-3) offers relatively easy access simply by lifting its cover plate to change or repair main or feeder telephone cable pairs. It is easy to locate because it is placed close to the house and sticks up out of the ground. Unfortunately, since it is in plain view, the cover can easily be removed and the cable pairs tampered with. During hot weather, the pedestal buried service terminal's cable pairs may actually bake inside the pedestal, causing the insulation material protecting the cable pairs to crack and fall off, resulting in wires

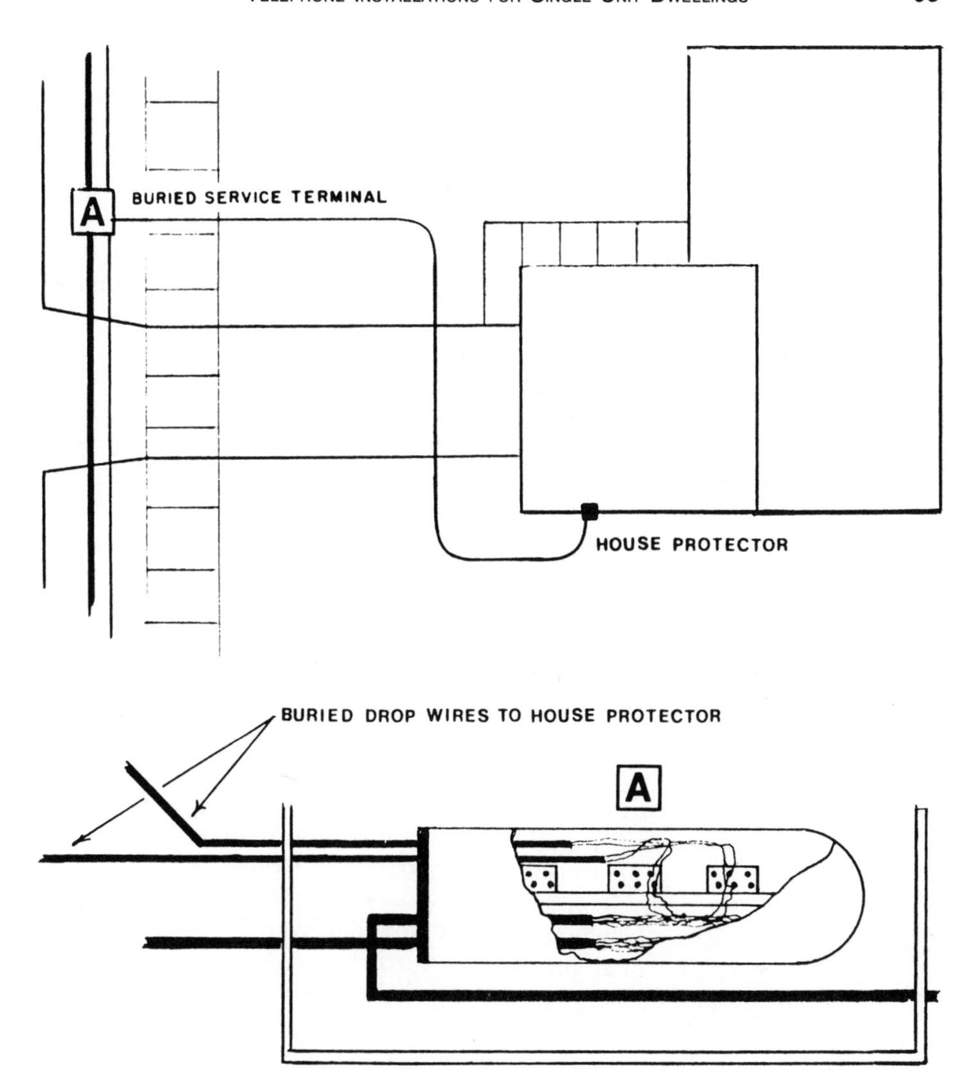

Fig. 3-2 Buried service terminal—hand box type.

that short, cross, or ground. Also, gophers and mice have been known to use pedestal service terminals as nesting grounds when they aren't chewing on the wires, all of which create telephone problems.

Drop Wire—Aerial Type

The aerial drop wire, the first connection for the station wiring, links your house protector to the unprotected service terminal (see Figure I-8 in the Introduction).

Because the aerial drop wire contains only one pair, a separate drop wire is needed for each telephone number you plan to have installed. There are, however, two exceptions: when the second telephone number is a party number or when the second phone number is built on line-

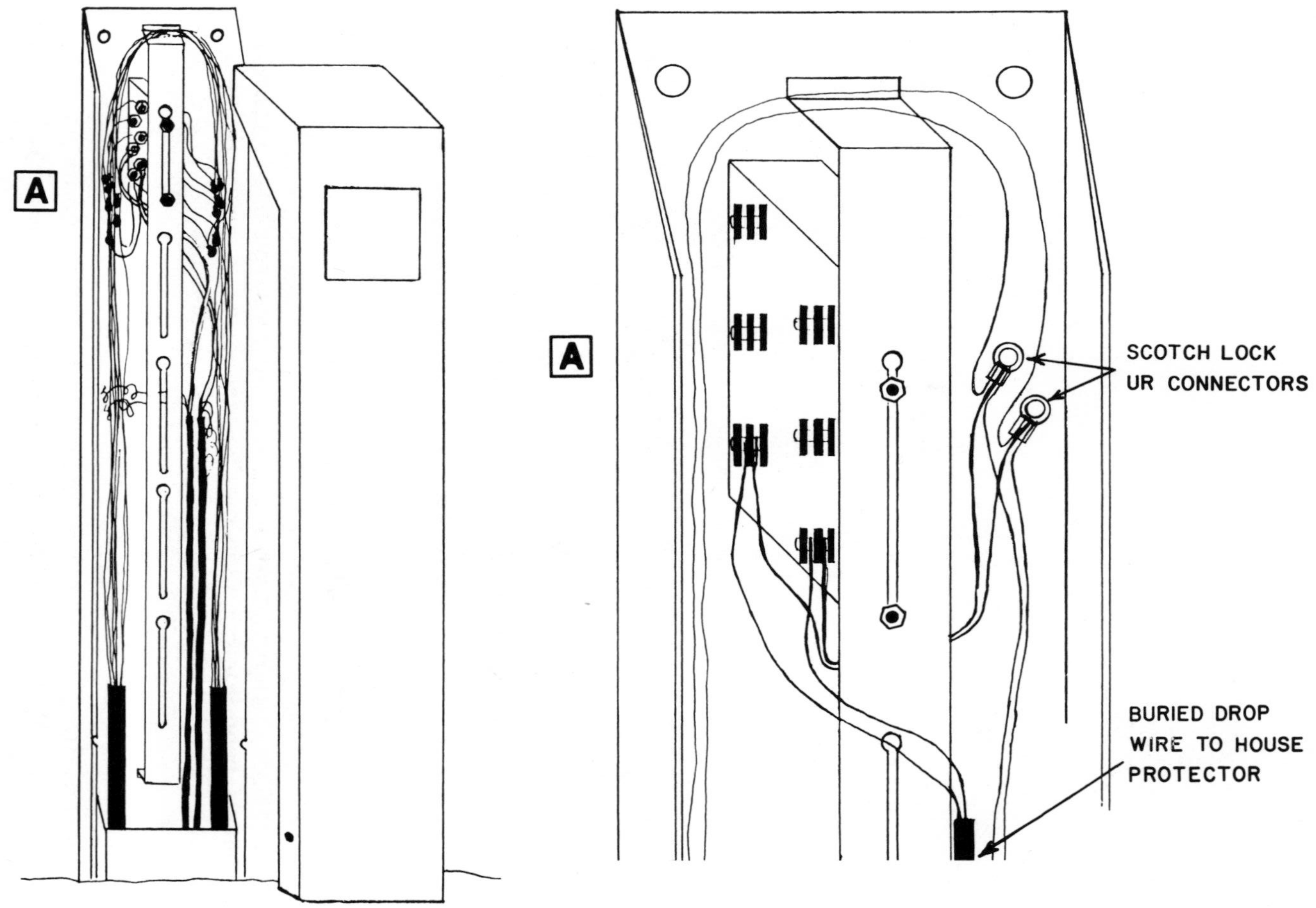

Fig. 3-3 Buried service terminal—pedestal type.

splitting equipment. (For more information, see "Second-party Lines" and "Line-Splitting Circuits" in Chapter Two.)

Sometimes an aerial drop wire crosses a driveway, street, or alley and if it is too low, it may be pulled out by passing vehicles (see Figure 3-1 for height specifications). If it is too close to a power line, it may pick up static or possibly connect with that line, causing damage to your telephone system. It is important for the aerial drop wire to have a slight sag, which allows a cushion against high winds that put undue stress on the wire and its connecting points.

To prevent a poor connection, the coating material on the aerial drop wire should be checked periodically for brittleness. Some telephone companies will repair damaged or unserviceable aerial drop wire at no expense to you, although this should be verified first through a quick call to the repair service.

Drop Wire—Buried Type

Buried drop wire, specifically designed to be placed underground, has extra layers of insulating material to keep out moisture (see Figure 3-46). When initially installed, the local telephone company usually places two buried drop wires with three pairs in each, resulting in enough pairs to provide up to six telephone circuits. Generally, it is located anywhere from 16 to 36 inches under the ground. If you intend to dig in your yard, chances are great that you'll damage the wire with the shovel. (Figure I-9 in the Introduction shows one possible route buried drop wire might follow.) The local telephone company will usually not charge to locate your buried drop wire but will almost always require advance notice.

If your area has a lot of gophers, it would be wise to put a very strong poison into the trench along with the newly installed buried drop wire to discourage the rodents from chewing on it.

House Protector

The purpose of the house protector is to protect you, your telephone, and your station wiring from damage caused by high voltage that may come in contact with your telephone system. It also serves as a general location where all inside wires connect to your drop wire.

There are three basic types of house protectors: the all-weather fused (see Figure I-10 in the Introduction and Figure 3-4), the indoor fused (Figure 3-5), and the indoor fuseless (Figure 3-6).

The all-weather fused house protector, the most common type, is designed to be placed just about anywhere inside or outside your home (see Figure I-10 in the Introduction and Figure 3-4). Referred to as a multistation house protector, it is available in different sizes to accommodate up to six drop wires at any one time. The multistation house protector is recommended over the single house protector, which may require two or more unsightly drop wires.

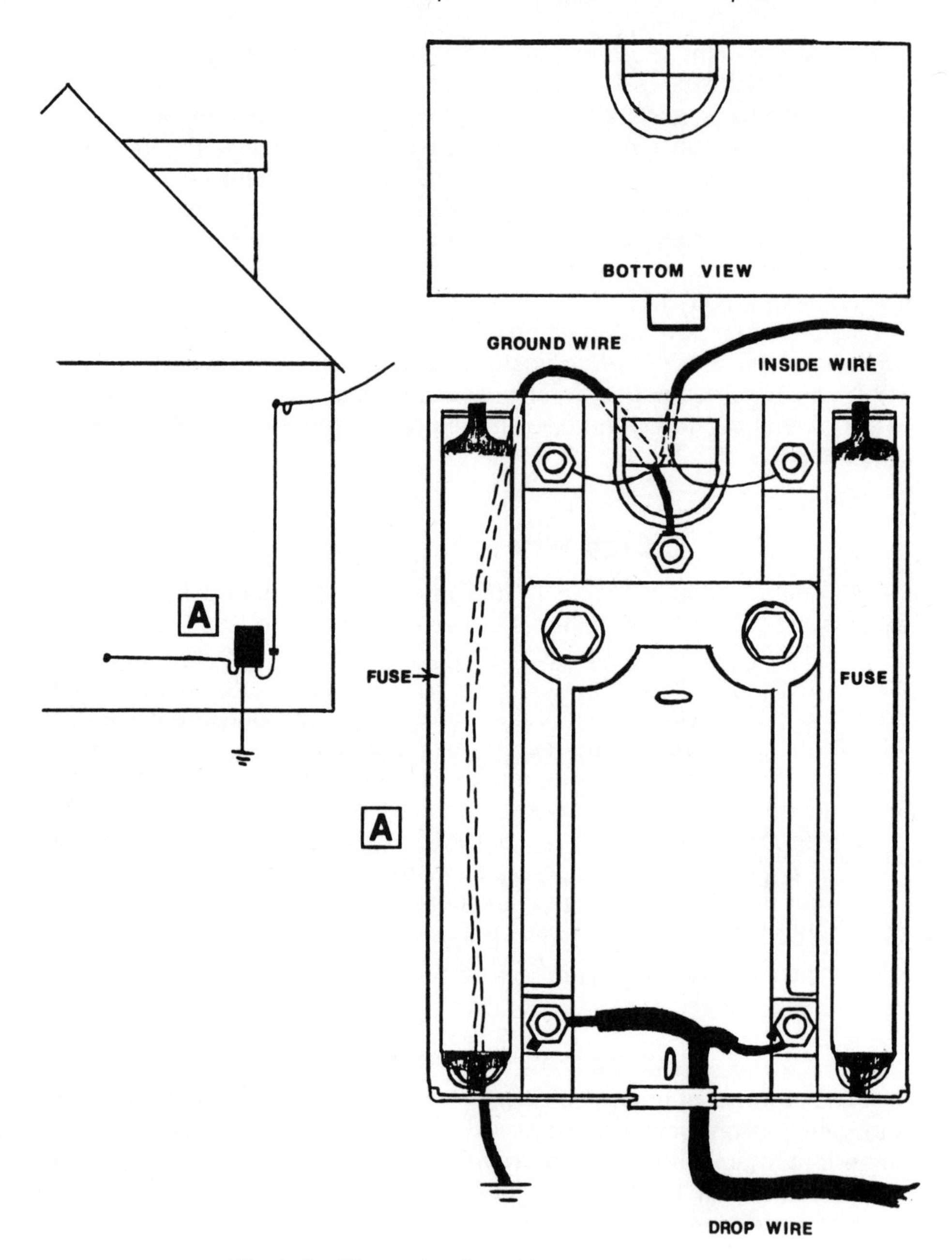

Fig. 3-4 All-weather fused house protector.

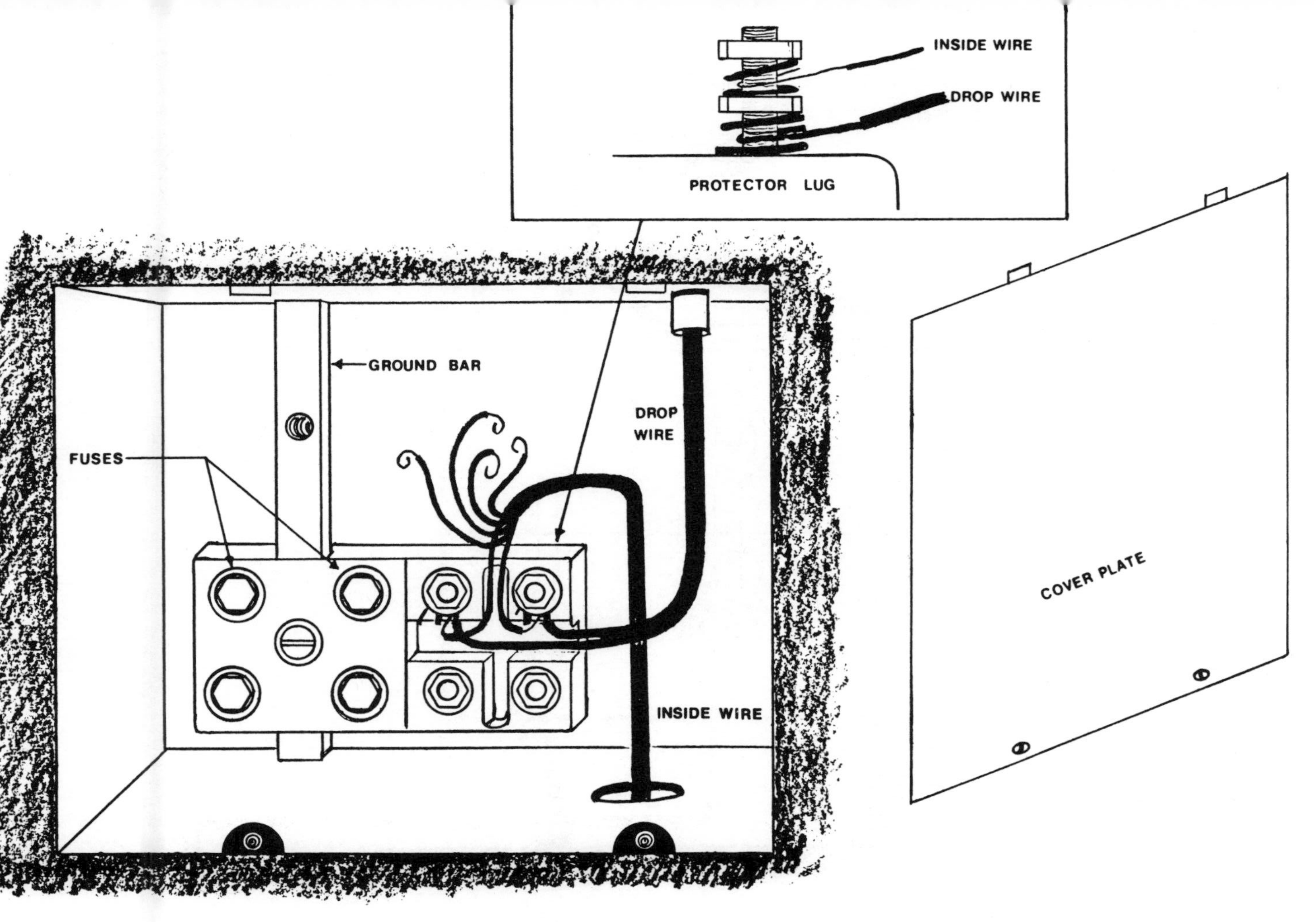

Fig. 3-5 Indoor fused house protector.

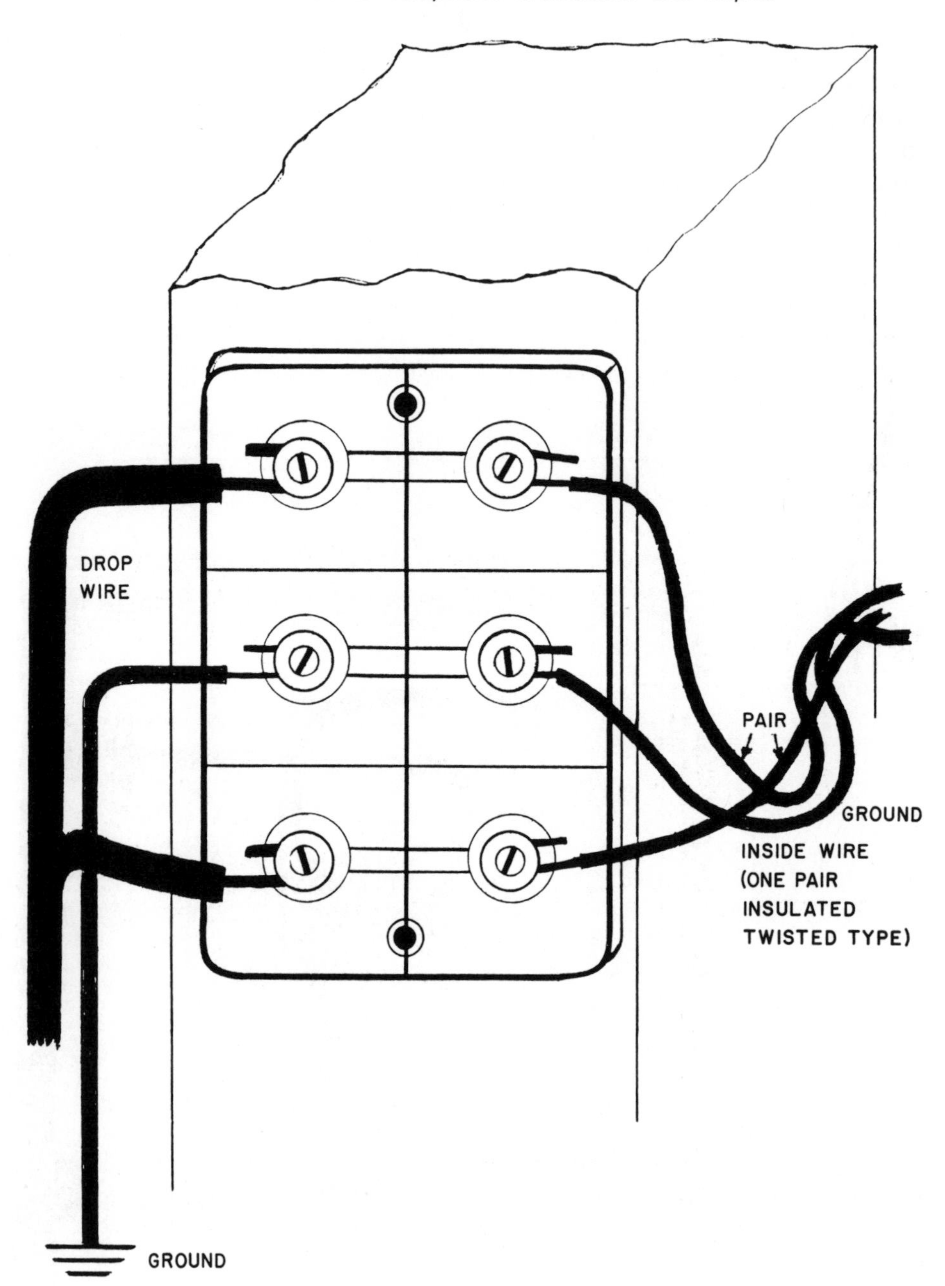

Fig. 3-6 Fuseless indoor house protector.

The indoor fused house protector, designed to be inside for protection from weather (see Figure 3-5), is generally installed during the initial construction of a house, when station wiring is prewired. Often, adding a room to your house may require the relocation of the drop wire and house protector, so you may want to consider adding a built-in type. Conduit and space will need to be drawn into the building plans, but most building contractors are familiar with the type of box required for this type of house protector.

The indoor fuseless house protector generally found in older homes, does not conform to current standards of protection. It is designed for protection in cases where the telephone itself is a part of the circuit (see Figure 3-6). This type of house protector has no fuses to disconnect the station wiring from the telephone system should high electrical voltage come in contact with it. Instead, it uses a ground wire within the inside wire to dissipate high electrical voltages that travel through your telephone and then back to the protector and the earth ground.

Many telephones today cannot handle this type of protection circuit. Some telephone companies may replace the indoor fuseless house protector at no expense to you, but again, a quick call to repair service would be wise to verify possible expenses to you.

Although the decision concerning the placement of your house protector is usually left up to the telephone company, you may request that it be located in an area with easy and safe access. While the attic is one common location, it is a poor choice. Most attics are hot, stuffy, and dirty and have little, if any, lighting. Also, many attics are now being sprayed with insulation material that not only hides the house protector but also is dangerous to breathe. Since telephone companies do not require their service personnel to crawl through areas that are unhealthy, access to your house protector may not be possible. In addition, when damp, the insulation material may short out your telephone system.

Another common house protector location—under the house—is generally used when crawl space is available, but this requires a serviceperson to lie on his or her back and work upside down. If the crawl space is muddy or unsanitary, servicepeople are under no obligation to enter and will often refuse. Snow, too, can hinder access.

Placement of the house protector under the eave of the roof is another common location, although a ladder will be needed for access. This location is especially acceptable if your yard is not secure; the house protector is placed out of reach from those who may wish to tamper with it. On an exterior wall about two or three feet up from the ground is a suitable location as it allows easy and safe access. If the occasional unusual location is difficult to find, your neighbor may be able to show you where the house protector is hidden.

A large number of telephone problems are associated with a house protector that is in poor condition. Dampness creates corrosion, which may require cutting off the corroded wires, stripping back the insulation, and replacing them. Corrosion might also cause the fuses to go bad, which creates static. To replace fuses, contact the telephone company since they are not sold on the open market. (Checking fuses will be covered in more detail in Chapter Five.) To control dampness, wipe the inside dry and make sure the cover to the house protector fits securely.

Spiders love to live inside house protectors, building their webs across the terminal lugs inside. In damp weather, spider webs may short out a pair. A stiff brush will usually wipe out this problem. Loose terminal lugs also cause telephone problems, so double-check to make sure they are tight each time you open the house protector.

Inside Wire

The inside wire, the third connecting point within the station wiring for single-unit dwellings, provides a connection between the house protector and the phone jack.

Installing your own inside wire and phone jacks can be an easy task since all you are doing is extending a pair from the house protector to the phone jack. Before installing inside wire, it is necessary to determine what type of inside wire already exists, how it is placed, and where existing phone jacks are located.

The four types of inside wire in use today are: the one pair insulated twisted type, one pair double insulated type, two pair double insulated type, and three pair double insulated type. Tip and ring, mentioned in the following illustrations, are terms used to define the polarity of a pair, with tip the most positive of the two wires and ring the most negative. If the polarity is reversed, some touch tone telephone instruments will not break dial tone, meaning the user will not be able to dial out; also, some rotary dialing telephone instruments will get bell taps when dialing. Figures 3-7a, 3-7b, and 3-7c show the inside wire color code for each pair, indicating which wire is tip and which is ring.

The one pair insulated twisted type consists of three wires of the same color, either white or brown (Figure 3–7a), one for ground and the other two for the pair. This type, with the ground wire included, is associated with the indoor fuseless house protector and its protection circuit. For today's type of installation, the ground wire is left unconnected, either cut or dangling free at the house protector and at the phone jack. Since this type of inside wire has only one pair, it will accommodate only one telephone number. An additional inside wire will need to be placed if you are planning to add another number. Although this type of inside wire is still in use in many older homes, it is uncommon for new installations.

The one pair double insulated type also consists of three wires (see Figure 3-7a). By insulating each individual wire and then jacketing the three wires with still another insulation layer, the arrangement allows for better protection against the elements. Each wire is color coded: green and red for the pair and yellow for the ground wire. This type handles only one telephone number and is now commonly used for installation.

The two pair double insulated type of inside wire is made up of four wires to accommodate two separate telephone numbers (see Figure 3-7b). The color code arrangement for this type of inside wire is green-red for the first pair and yellow-black for the second. This is the most popular inside wire now being sold at telephone stores.

The three pair double insulated type is made up of six wires (three pairs) to handle three separate telephone numbers (see Figure 3-7c). This type is found with two arrangements of color code: one is where the first pair is blue–blue-white, the second orange–orange-white, and the third green–green-white; the other is where the first pair is red-green, the second yellow-black, and the third blue–blue-white. Although this is the standard color code for wiring phone jacks, adherence to it is not always possible, which will be explained later in this chapter.

The installation of the three pair double insulated inside wire is preferred because it allows for options in expanding your home telephone system without the need to add extra inside wire.

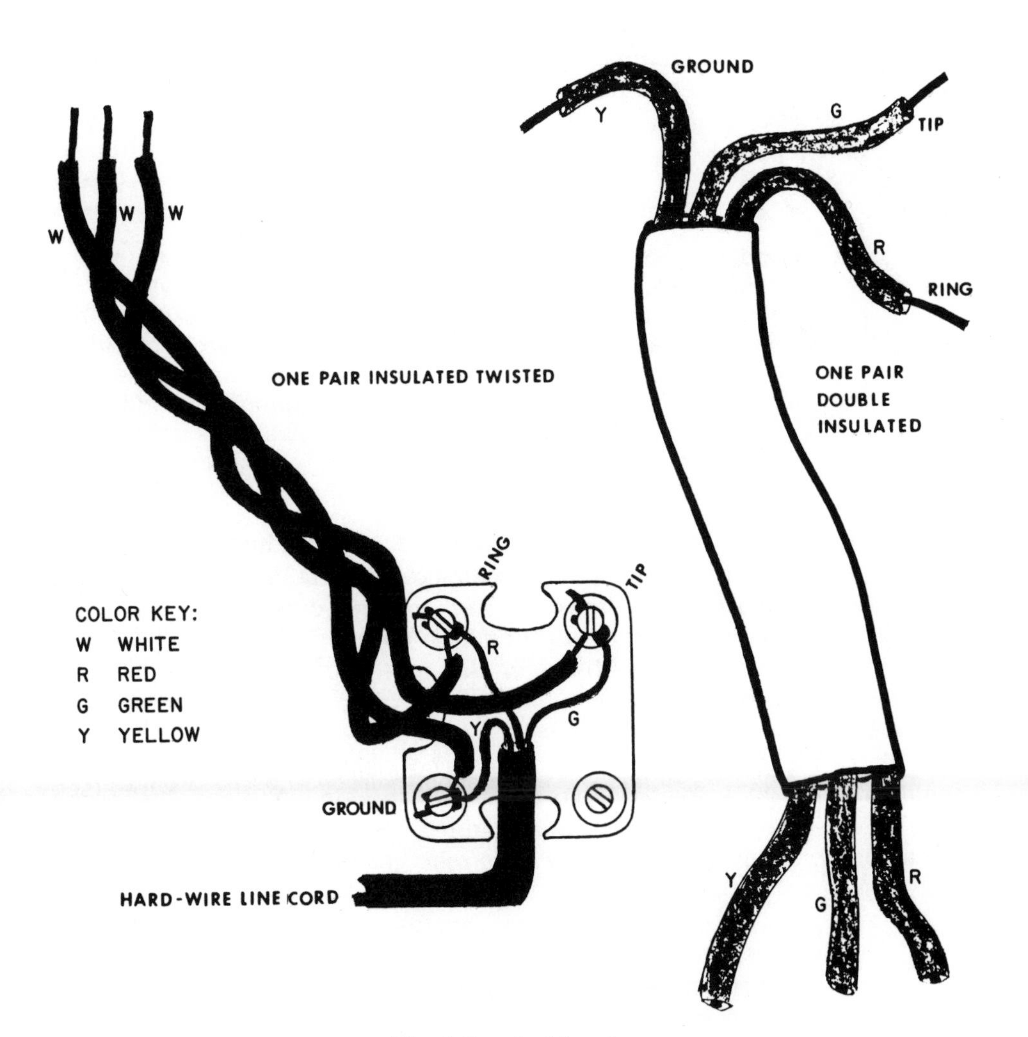

Fig. 3-7a Inside wire.

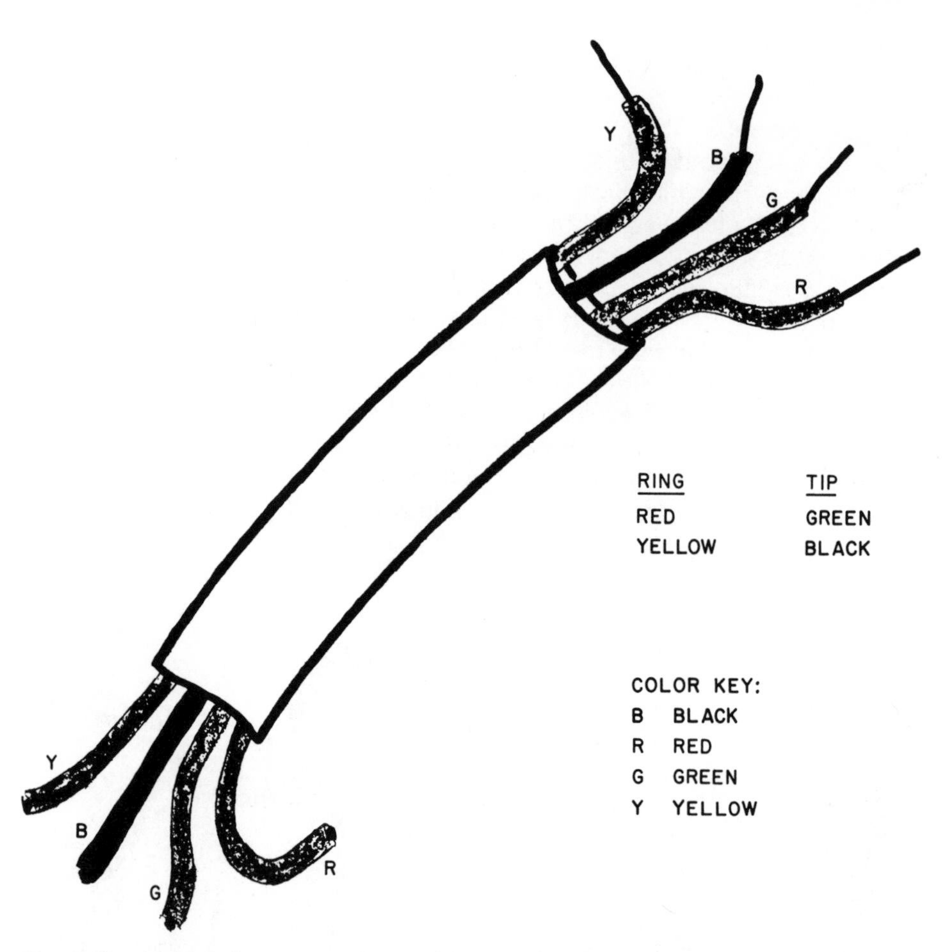

Fig. 3-7b Inside wire—two pair double insulated. This is the most popular inside wire found at phone stores.

Because of the deregulation of the telephone industry, telephone companies are not responsible for repairing inside wires and phone modular connectors installed by the customer or other companies, and they often impose a penalty charge (around $55) for such repairs.

If the inside wires and phone modular connectors in your home were not installed by the telephone company, you will need to have your telephone company install an *interface phone modular connector.* The interface phone modular connector serves as a connection point, demarcation point, and test point for both you and the telephone company. The interface PMC requires a phone modular connector plug to be installed at the house protector or protected service terminal end of your inside wire.

Methods of Installing Inside Wires

There are three common methods for installing inside wire: the prewire method, run-under method, and wrap method.

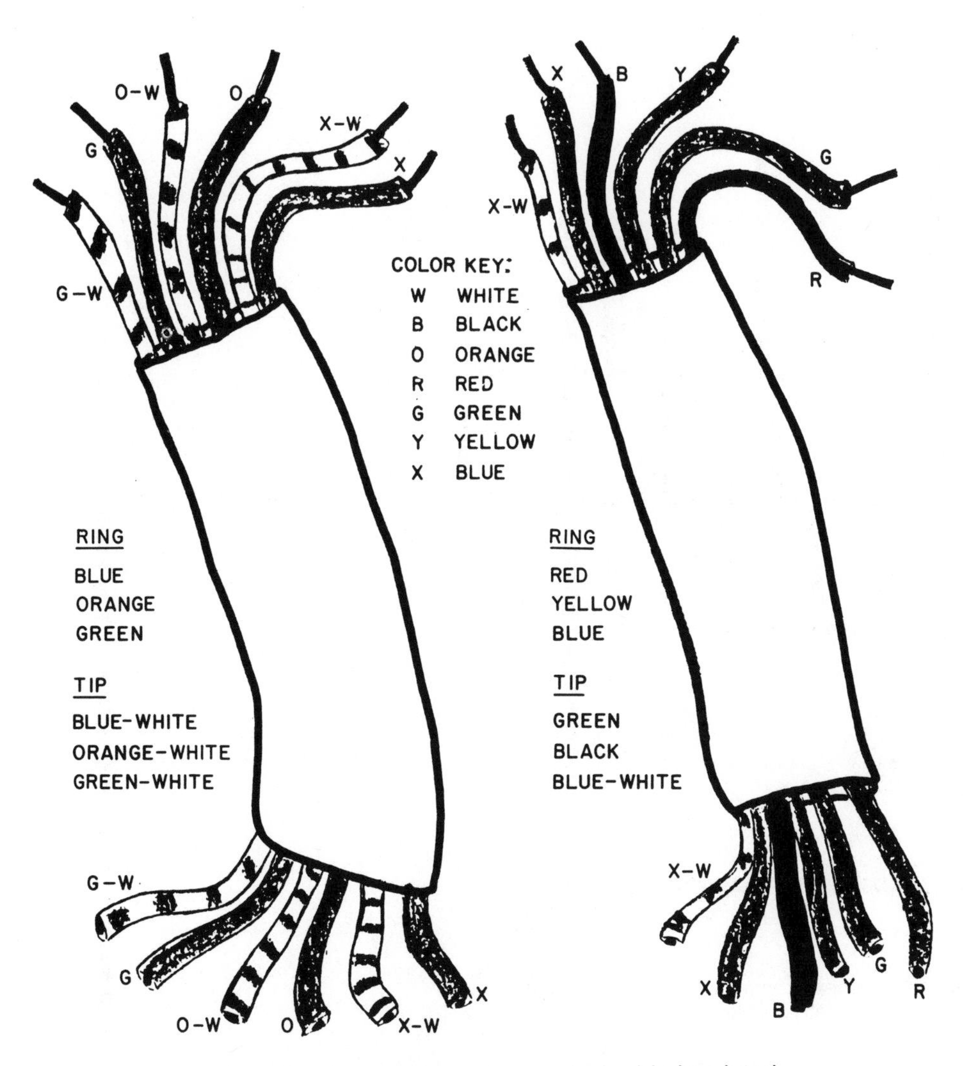

Fig. 3-7c Inside wire—three pair double insulated.

Prewire Method

The prewire method of installation (Figure 3-8), generally found in newly constructed homes, is where the inside wire is placed inside the walls during the framing stages of construction; thus, wall fixtures are placed so that flush-mount jacks can be installed, as demonstrated in Figure 3-9. This method is recommended because it is more attractive than having inside wire running along baseboards or fastened to exterior walls. If you are planning any room additions to your home, make arrangements to prewire the inside wire while you are at the framing stages of construction. A disadvantage is that if prewired inside wire should go bad, there is no way to replace it without tearing down the wall.

Proper installation of the prewired inside wire is important: if it is nicked, breaking the insulation, dampness may either short the pair or sever it open; also, if pinched while being fastened to a board within the frame of the wall, shorting or severing may occur in the pair. (These problems may not show up for months or even years after the installation.)

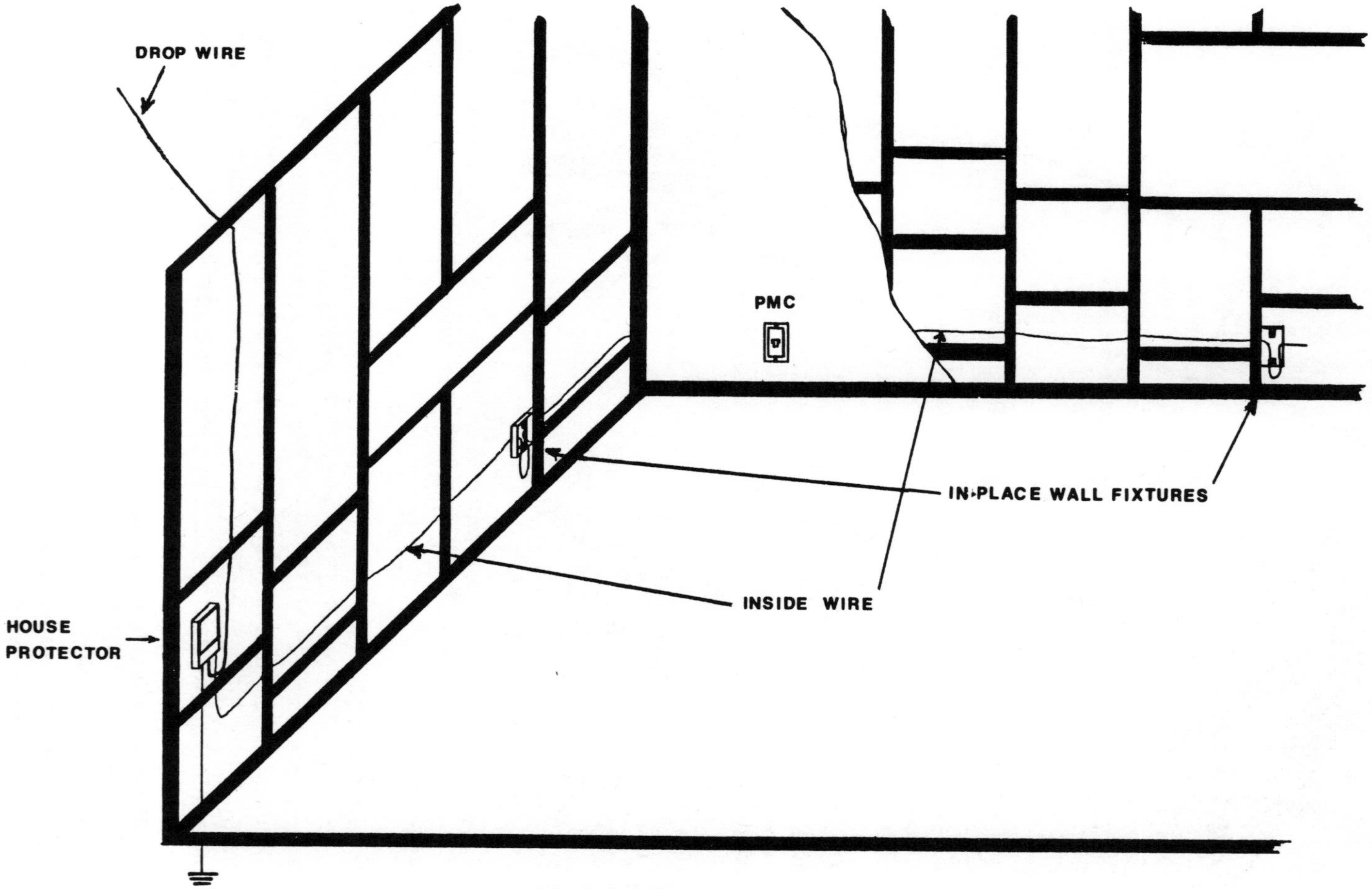

Fig. 3-8 Prewire method.

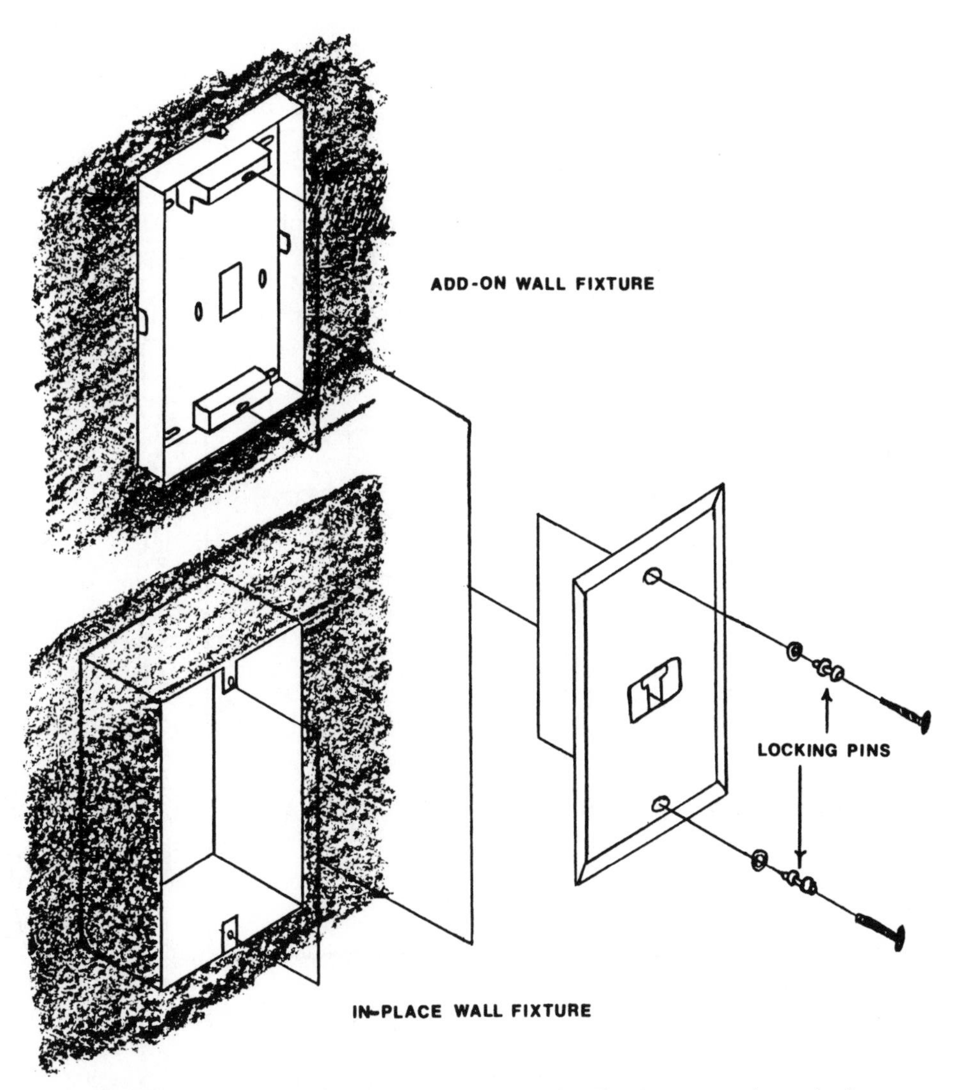

Fig. 3-9 Phone modular connector (PMC): Flush-mount phone jack.

The three pair double insulated inside wire is often an advantage in this case since sometimes when a pair goes bad in a wall, it is possible to switch it with another pair within the three pair double insulated inside wire. This is one explanation, as mentioned earlier, as to why strict adherence to a standard color code with inside wire is not always possible.

Run-Under Method

The run-under method (Figure 3-10), generally used when a house has a crawl space, allows the inside wire to be run under the floor and brought up through the floor next to an interior wall close to the baseboard.

In Figure 3-10, the inside wire is shown hanging down only for clarity, since this is an incorrect method for attachment. The inside wire should be

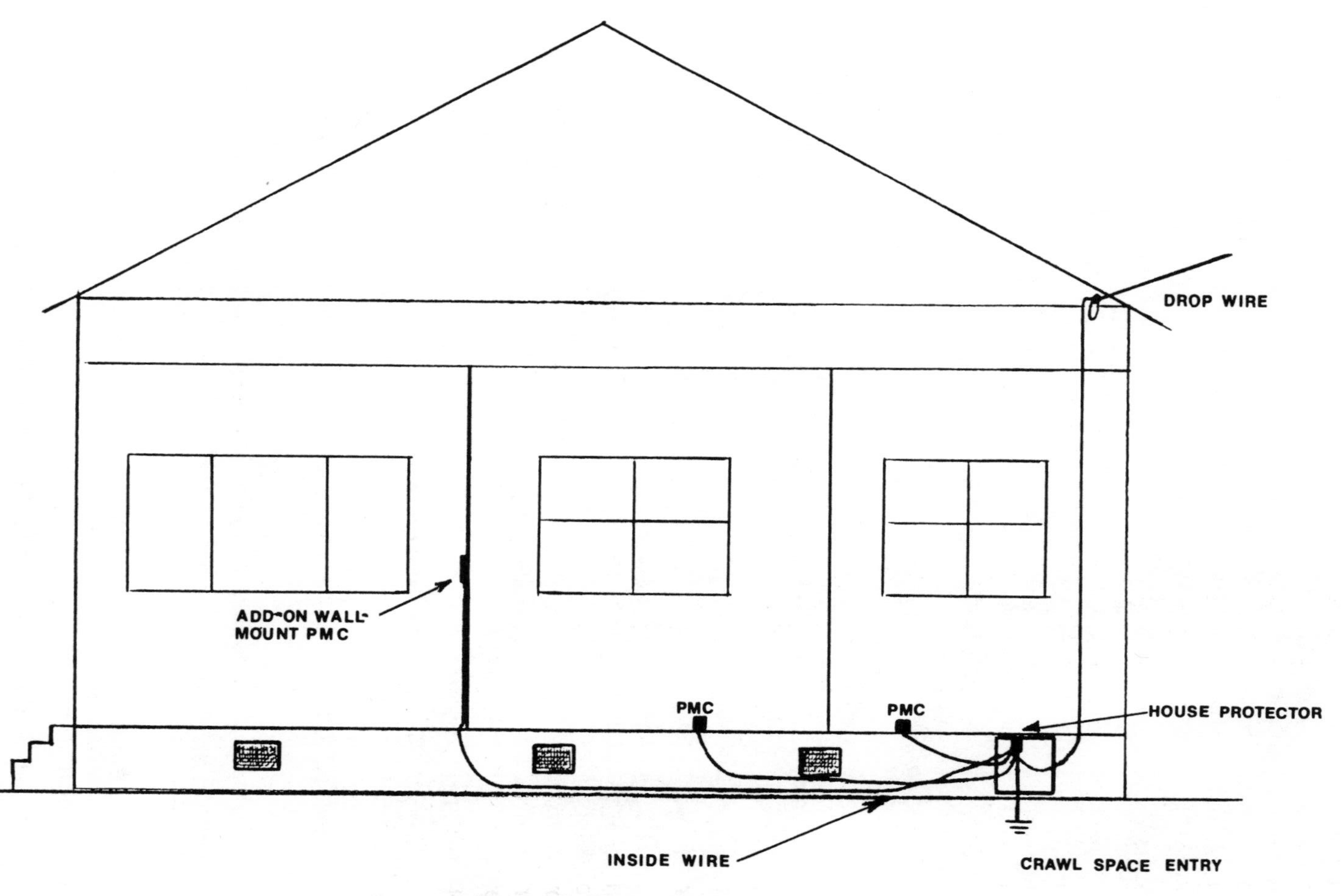

Fig. 3-10 Run-under method.

attached up off the ground; if left on the ground, a serviceperson may accidentally pull it out. (The telephone repairperson is not the only one involved in these cases; often, plumbing, heating, and electrical service-people need access to house crawl spaces because their connections are also located underneath the house.) During the rainy season, dampness may create corrosion if the inside wire is left hanging on the ground.

Wrap Method

With the wrap method (see Figure I-11 in the Introduction), the inside wire is attached to an exterior wall and then brought through that wall to the inside of the house. Figure 3-11 illustrates two ways to wrap the inside wire around a house when the desired location for a phone jack is on the opposite side. Attaching the inside wire to the lip of the eave is recommended, if possible, since less inside wire is required to go the distance, as compared to using the steplike slats that support the eave.

Circuitry for Connecting Inside Wire to Phone Jacks

There are two basic circuits used to connect the inside wire to the phone jack: the in-series circuit and the home-run circuit. Both have advantages, and in many homes, both are used together.

In-Series Hookup

The initial advantage to the in-series hookup (Figure 3-12) is that less inside wire is used to connect the phone jacks; but if it becomes necessary to isolate a telephone problem, it may require the inspection of just about every phone jack in the house. If the in-series hookup has only one pair of inside wire, options for adding additional telephone numbers are nonexistent without first having to place additional inside wire.

Home-Run Hookup

One advantage to the home-run hookup (Figure 3-13) is that isolating a telephone problem can be done easily at one location—the house protector—by disconnecting each individual inside wire until the problem clears. This type of hookup allows you to add additional telephone numbers, no matter if the inside wire is one pair or three. Since more inside wire is required, determining which inside wire goes to which jack might prove difficult; however, this problem can be quickly figured out by checking each inside wire from the house protector to see which jack goes dead. (More about this subject will be covered in Chapter Five.)

Attaching Inside Wire to Different Surfaces

When installing a phone modular connector, you should consider what route the inside wire will run so that you can determine what type of surface is necessary for attachment.

Wood

Wood is the easiest and most common surface to contend with and, for this reason, should be given primary consideration when planning an

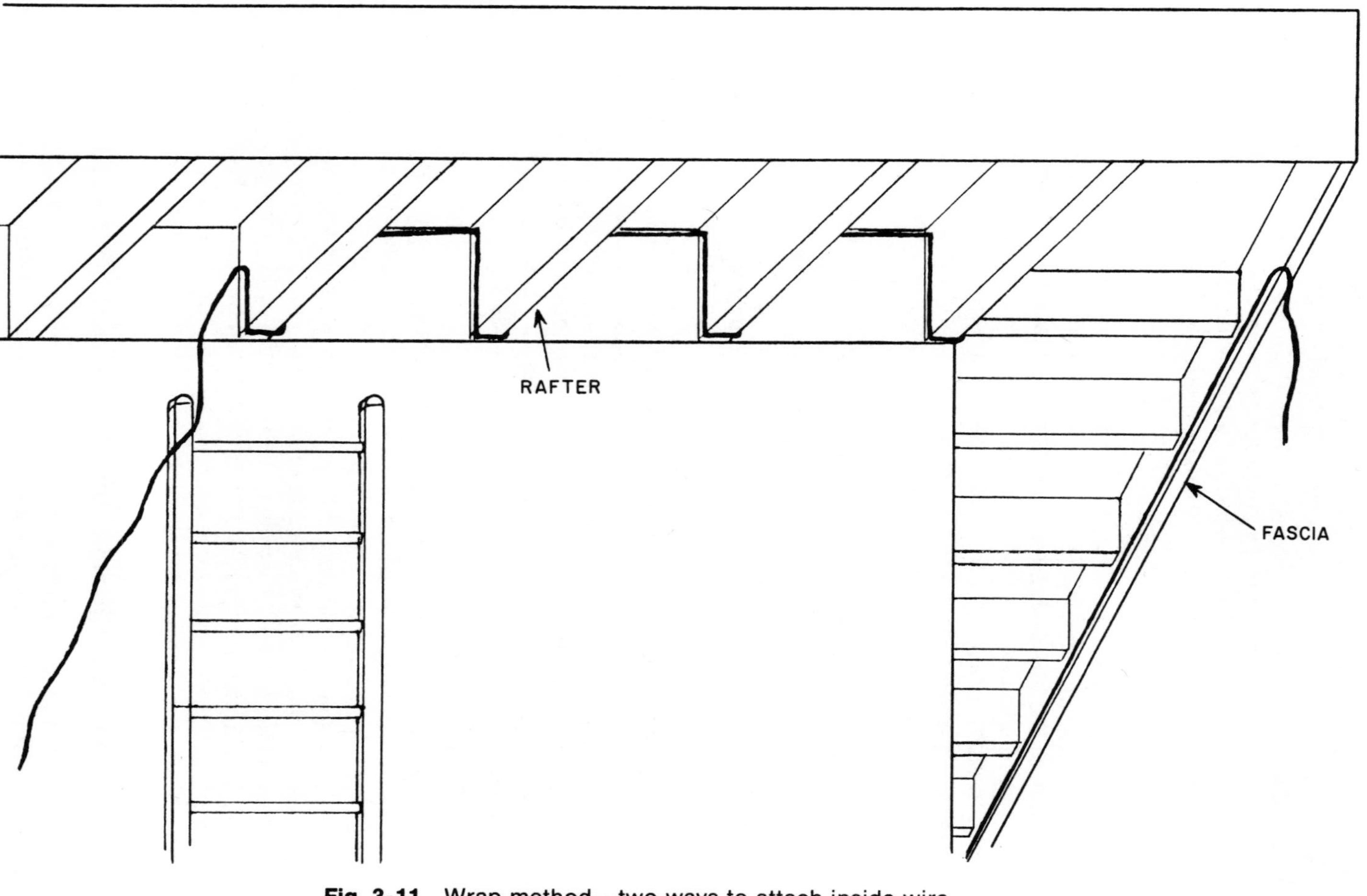

Fig. 3-11 Wrap method—two ways to attach inside wire.

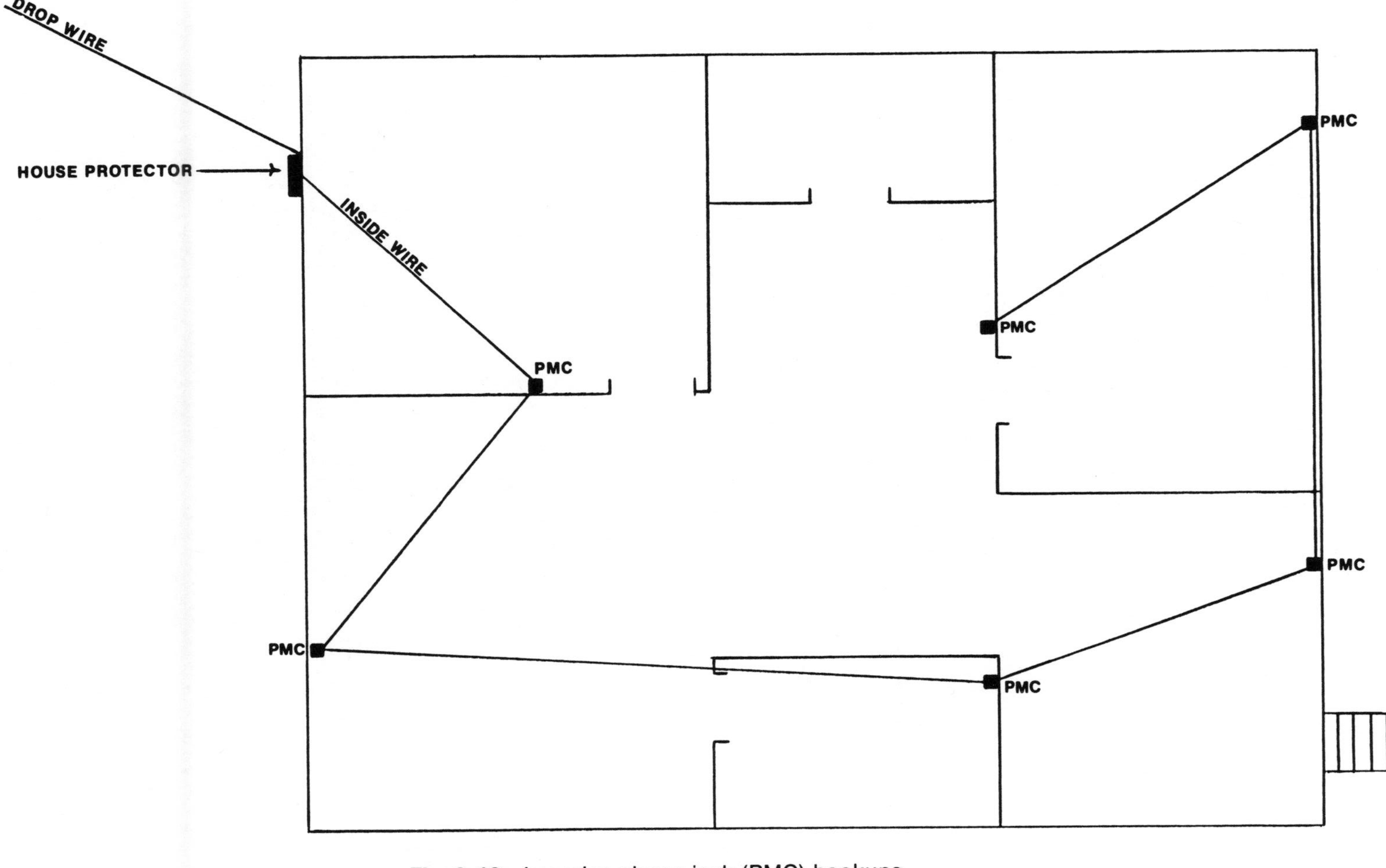

Fig. 3-12 In-series phone jack (PMC) hookups.

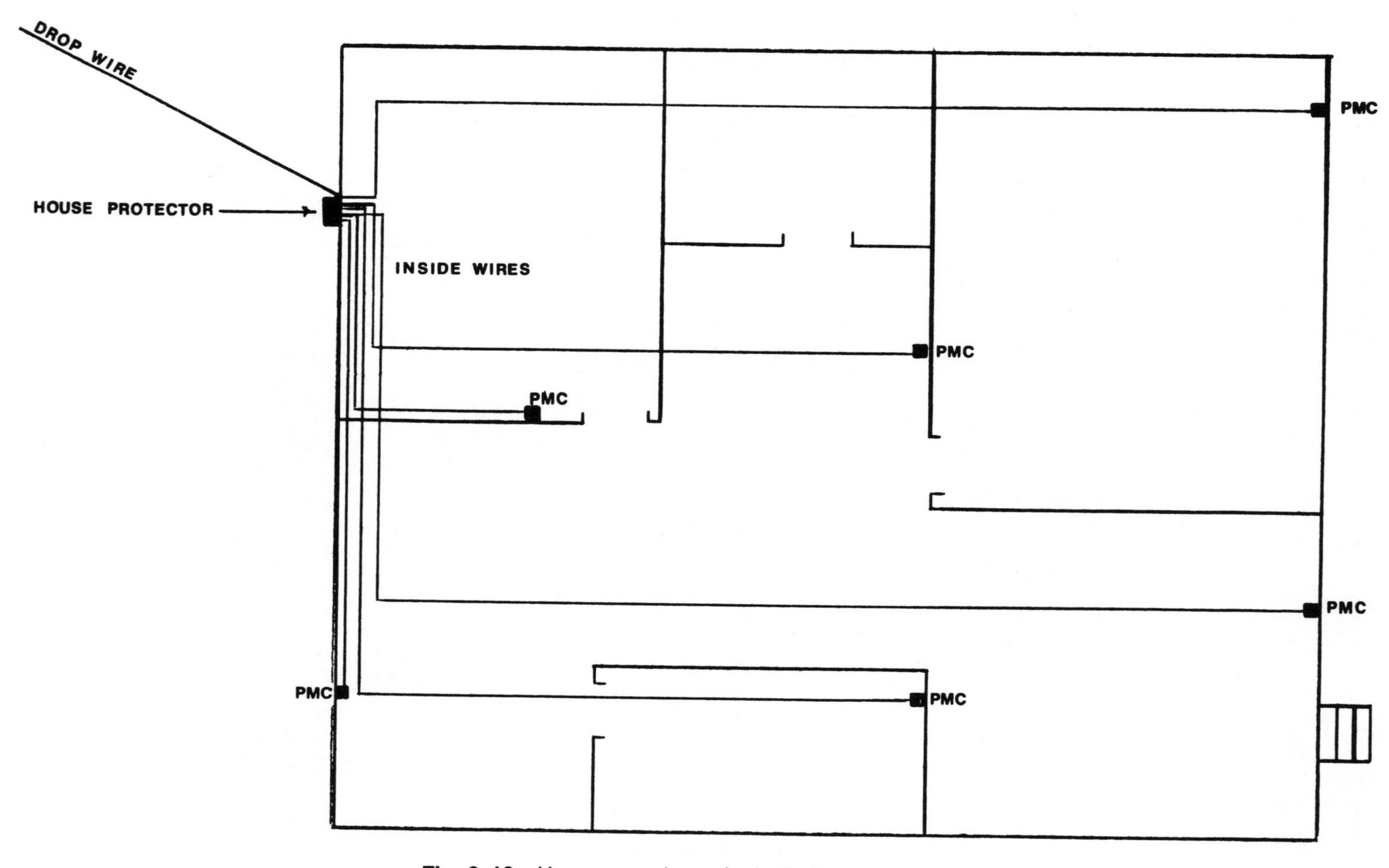

Fig. 3-13 Home-run phone jack (PMC) hookups.

installation. A simple way to attach inside wire to wood is to use small, insulated staples that are available at home centers. These are easier to work with than screws and clamps and less expensive than a special staple gun.

The most popular tool used by telephone installation personnel for attaching inside wire to wood is the T-25 staple gun with T-25 staples (Figure 3-14). If you feel the staple gun (around $30) is too expensive, check your local telephone store because some are renting the staple gun for just a few dollars. If you do use the staple gun, be careful not to shoot a staple into the inside wire; it may short or open the newly installed telephone circuit.

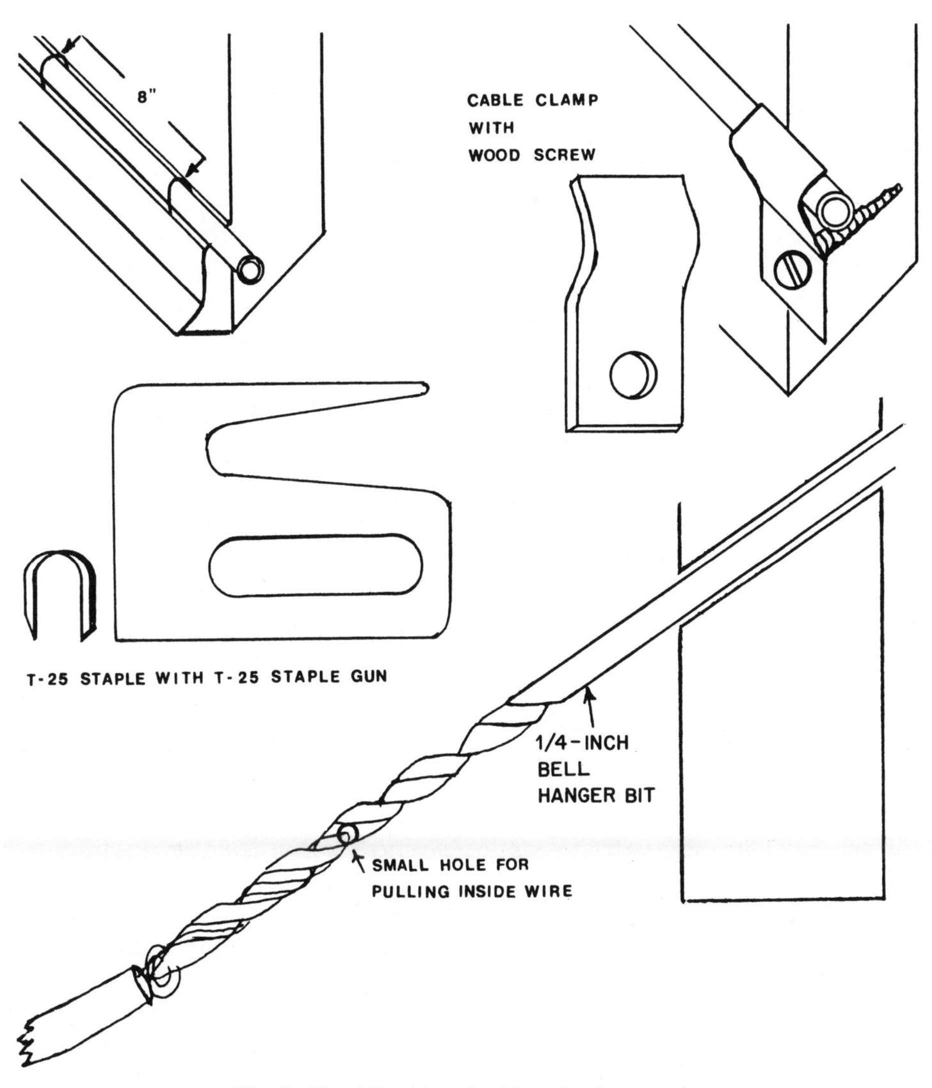

Fig. 3-14 Attaching inside wire to wood.

An alternative to the staple gun is the inside wiring cable clamp with wood screws (see Figure 3-14).

A quarter-inch diameter, extra long drill bit called a bell hanger bit is perfect for passing inside wire through wood or masonry walls. The small hole drilled through the bell hanger bit allows you to pull the inside wire through a hole drilled in the wall (see Figure 3-14). Notice that in an exterior wall, such as the one in the figure, the hole is drilled at an angle so that moisture from the outside will not seep into the wall or house. (The angle is not necessary when drilling through interior walls.)

If you don't have or can't find a bell hanger bit, an ordinary wood or masonry bit can be used to drill the hole. Then lightly tape the inside wire to the end of a piece of coat-hanger wire, and push the coat-hanger wire through the hole. Go to the other side of the wall and pull the phone wire through—just like a needle and thread.

Masonry

When you're attaching inside wire to masonry, cable clamps will be needed. Short masonry nails or drive pins will fasten the clamps to concrete blocks or soft bricks. Plastic wall anchors (Figure 3–15) can also be used, but they are more difficult to work with and require holes to be drilled. To place the wall anchors, use an extra long, quarter-inch masonry drill bit. The plastic wall anchors come in handy when attaching inside wire and phone modular connectors to dry wall.

Metal

When you're attaching inside wire to metal surfaces, cable clamps with sheet metal screws will be needed unless the metal surface is too hard, in which case, use tie wraps with adhesive tie wrap anchors (see Figure 3-15).

When attaching inside wire to any type of surface, be careful to avoid drilling into walls near electrical wiring or plumbing. Drill from the inside to the outside (again at a slight angle), and try to place the wire where it will not be pulled off or damaged. When anchoring inside wire with staples or cable clamps, be careful not to pinch the wire.

Running Inside Wire to Phone Modular Connector Locations

When running inside wire to phone modular connector locations, begin by reviewing the existing station wiring since it may be possible to utilize it. (Figures I-11 in the Introduction, 3-8, and 3-10 through 3-13 may help you determine what type of station wiring exists in your house.)

The easiest and most cost-saving method for installing a new phone modular connector for either an extension of an existing telephone number or the installation of a second one is to use one of the existing connectors as a splicing point for the new addition. Remove the phone jack cover plate. If there is more than one pair within the inside wire already connected, it will be possible to add a second number. The extra pair must

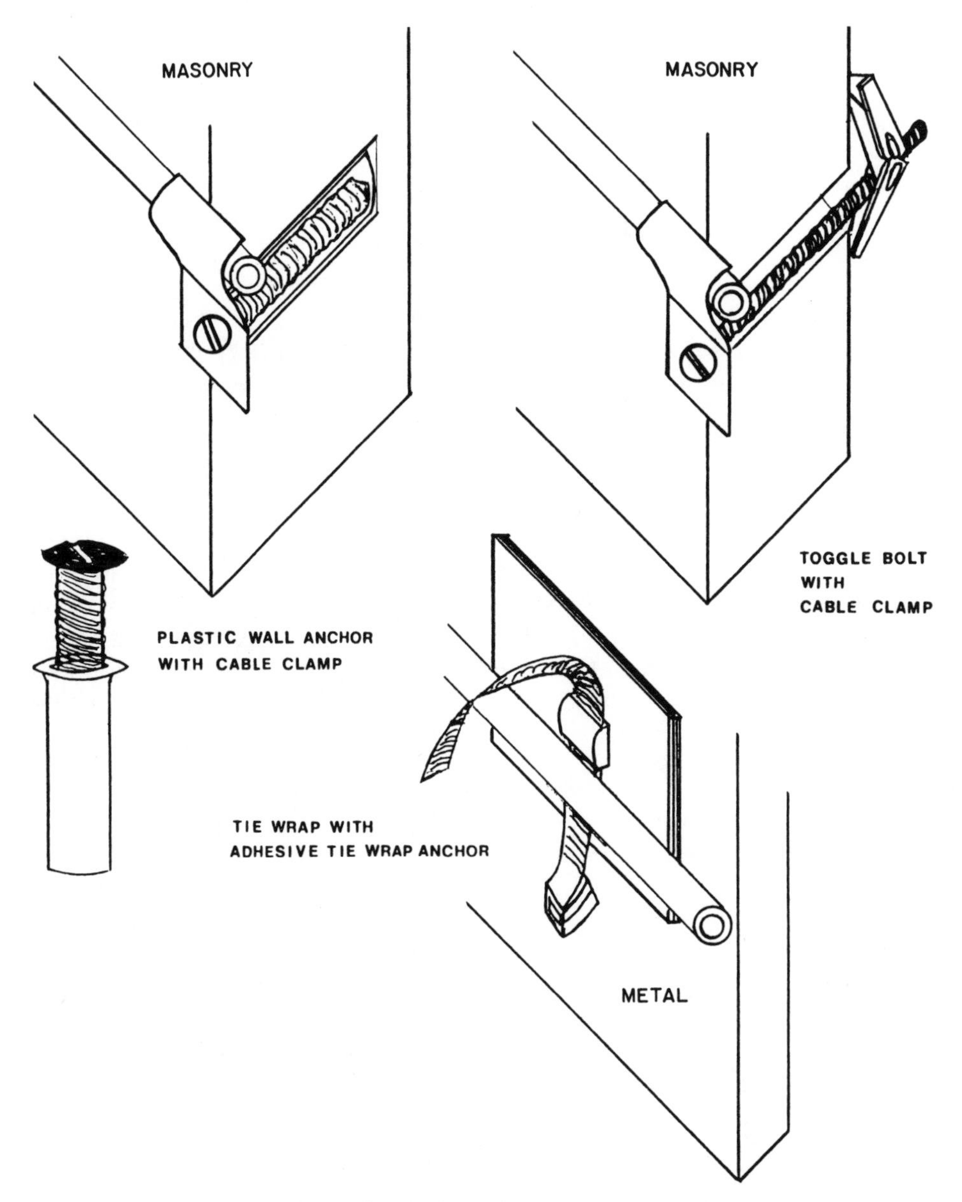

Fig. 3-15 Attaching inside wire to masonry and metal.

travel back to the house protector or protected service terminal; if it doesn't, tone it from the phone modular connector back to the house protector or protected service terminal to ensure that it does. (Chapter Five details how to tone inside wire.) If you are simply adding an extension of an existing telephone number, the only requirement is that the existing number be working from the phone modular connector you intend to use.

Figure 3-16 illustrates how an existing phone modular connector is utilized to install a new connector across a room. (Splicing techniques will

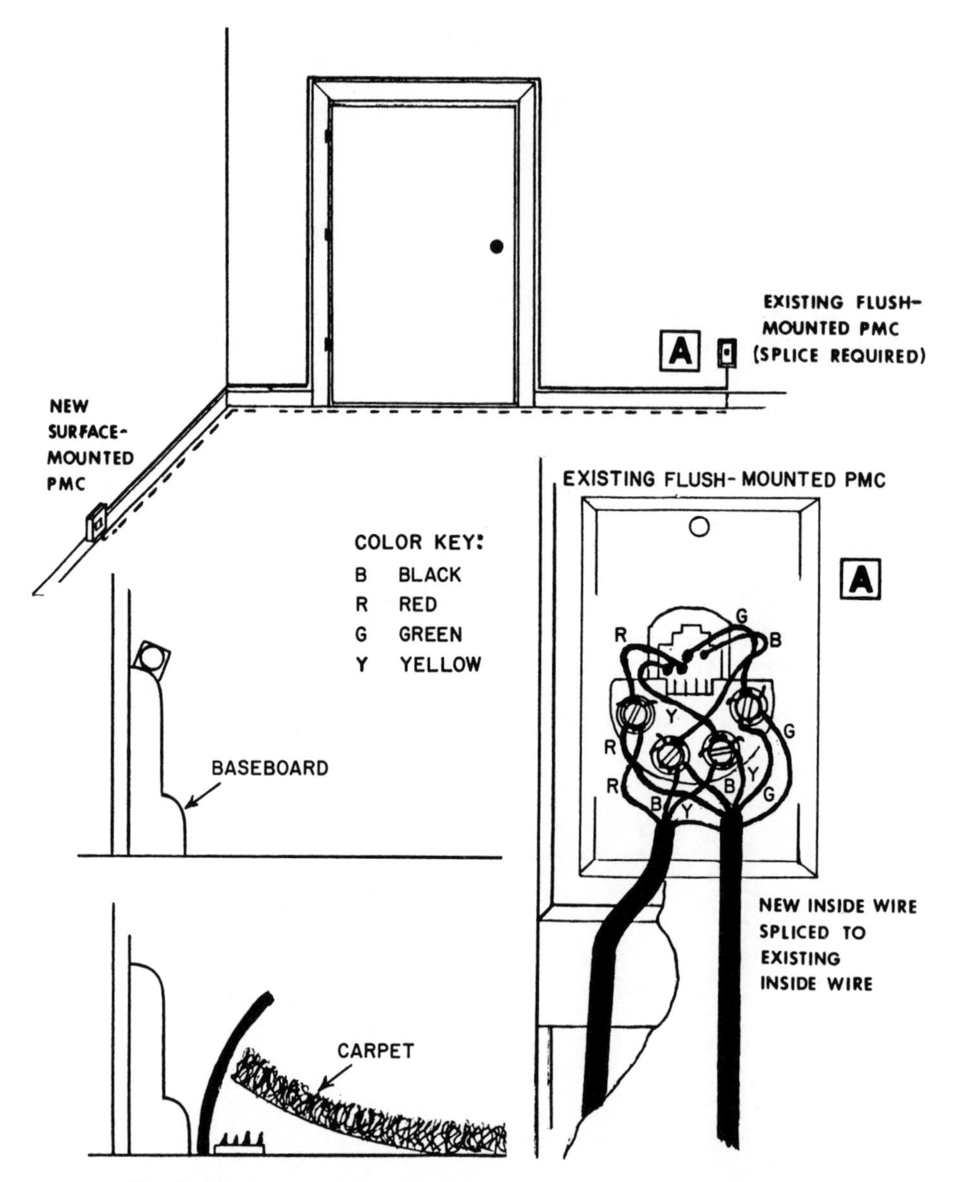

Fig. 3-16 Inside wire runs under carpet or along baseboard.

be covered in the following section.) There are two methods for reaching a location on the other side of a room: one uses a T-25 staple gun to staple inside wire to the baseboard and the doorway trim; the other hides the inside wire under the carpet, keeping clear of the carpet tack strip so nails won't puncture the inside wire.

Figure 3-17 illustrates five examples for running an inside wire around obstacles. When deciding where to place inside wire, consider crawl space first. Keep in mind that when running wire up walls, you can conceal it with wood molding, especially if it's in the corner of a room.

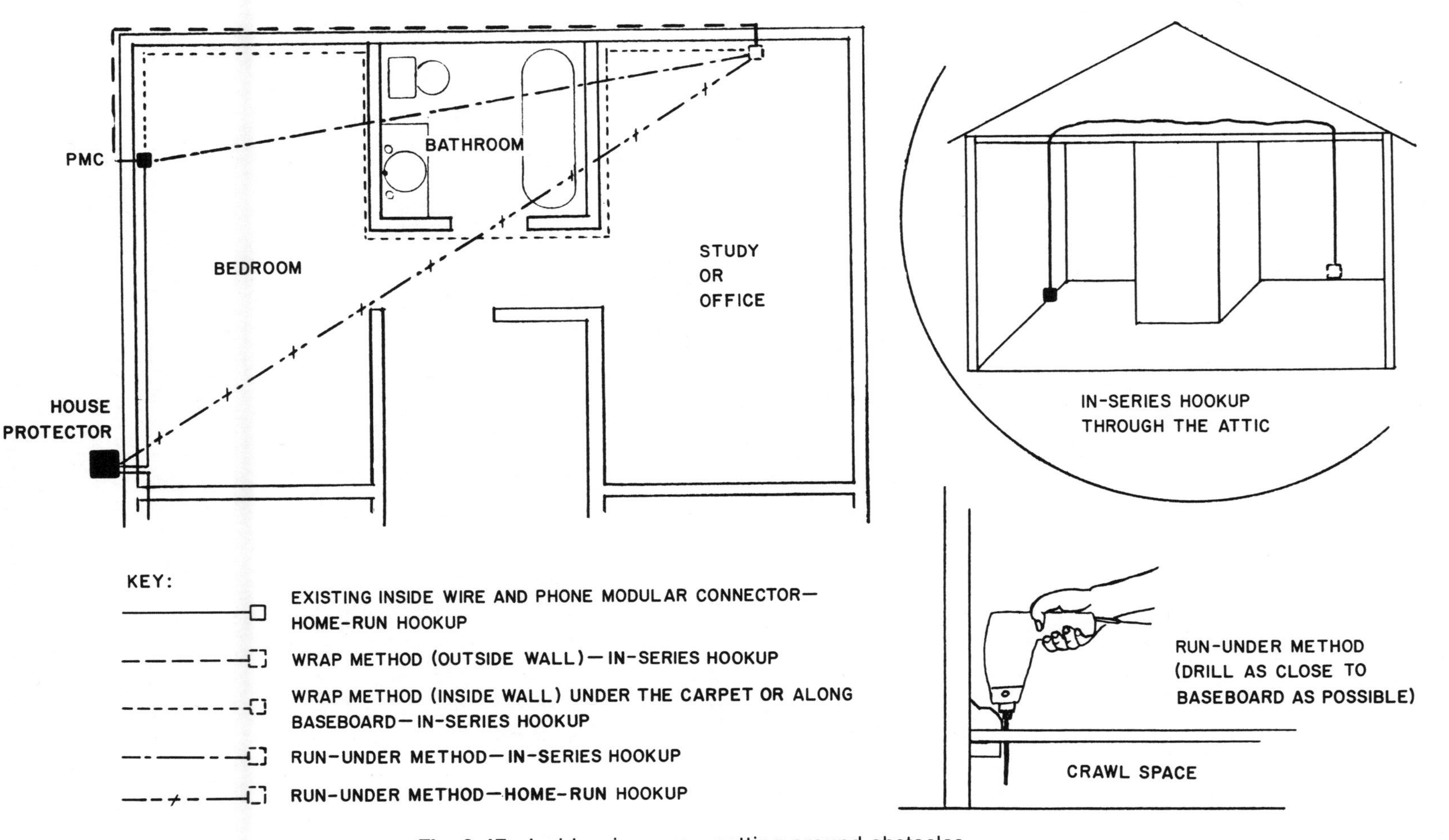

Fig. 3-17 Inside wire runs—getting around obstacles.

The wrap method, using an outside wall, is one technique for avoiding the bathroom (bathtub). Drill through the exterior wall at the existing phone modular location then attach the inside wire to the outside wall, running it around the house until you reach the room where the new phone modular connector is to be installed. Drill a hole and pull your inside wire in.

Another method illustrates using the wrap method on inside walls, where you can either attach it to the baseboard or hide it under the carpet. A third method, requiring a crawl space, uses the run-under method with an in-series hookup. This necessitates drilling a hole at the baseboard through the floor and into the crawl space. Drilling through the actual baseboard will sometimes split it; in addition, there is often a supporting beam directly under the baseboard, making it difficult to get a good hole through which to pull the inside wire. It is best, therefore, to drill as close to the baseboard as possible but not right through it.

A fourth way also uses the run-under method, but instead of connecting to an existing phone modular connector, the inside wire is connected to the house protector, making this a home-run hookup. The final method demonstrates how to use the attic in homes that have no crawl space. Take the inside wire up the wall into the attic, string it over to the location for the new phone modular connector, and then bring it down again.

Once the course your inside wire will run and the types of surfaces it will be attached to have been determined, you will need to know a few splicing techniques.

Splicing Inside Wire

By splicing inside wire, you will be able to add length to accommodate extension phone modular connectors. Figure 3-18, which details splicing inside wire at either connector blocks or phone modular connectors, shows the use of Scotch Lock UY Connectors, which are commonly used at the house protector or at locations where it is impossible to use the connector blocks or phone modular connectors. (Scotch Lock UY Connectors are available at most electronics or electrical stores.)

Never splice inside wire in locations where the connections might be exposed to water or moisture, including along the kitchen or bathroom floors. Splices left exposed to water or moisture will corrode and short or open the telephone circuit.

Figure 3-19 shows the proper method for stripping outer and inner insulation away from the inside wire and illustrates how to attach it to terminal lugs. The bare wire must be wrapped in the direction the terminal lug turns as it tightens; otherwise, when you tighten the lug, the wire will slip off. Also, each wire must have a spacer between it and another wire or when you tighten the lug, one wire will pinch and cut the other. Extra slack of approximately three inches allows for mistakes and changes you may wish to make in the future.

A workable technique for stripping inside wire is shown in Figure 3-20, using needlenose pliers with a cutting edge. After carefully scoring the

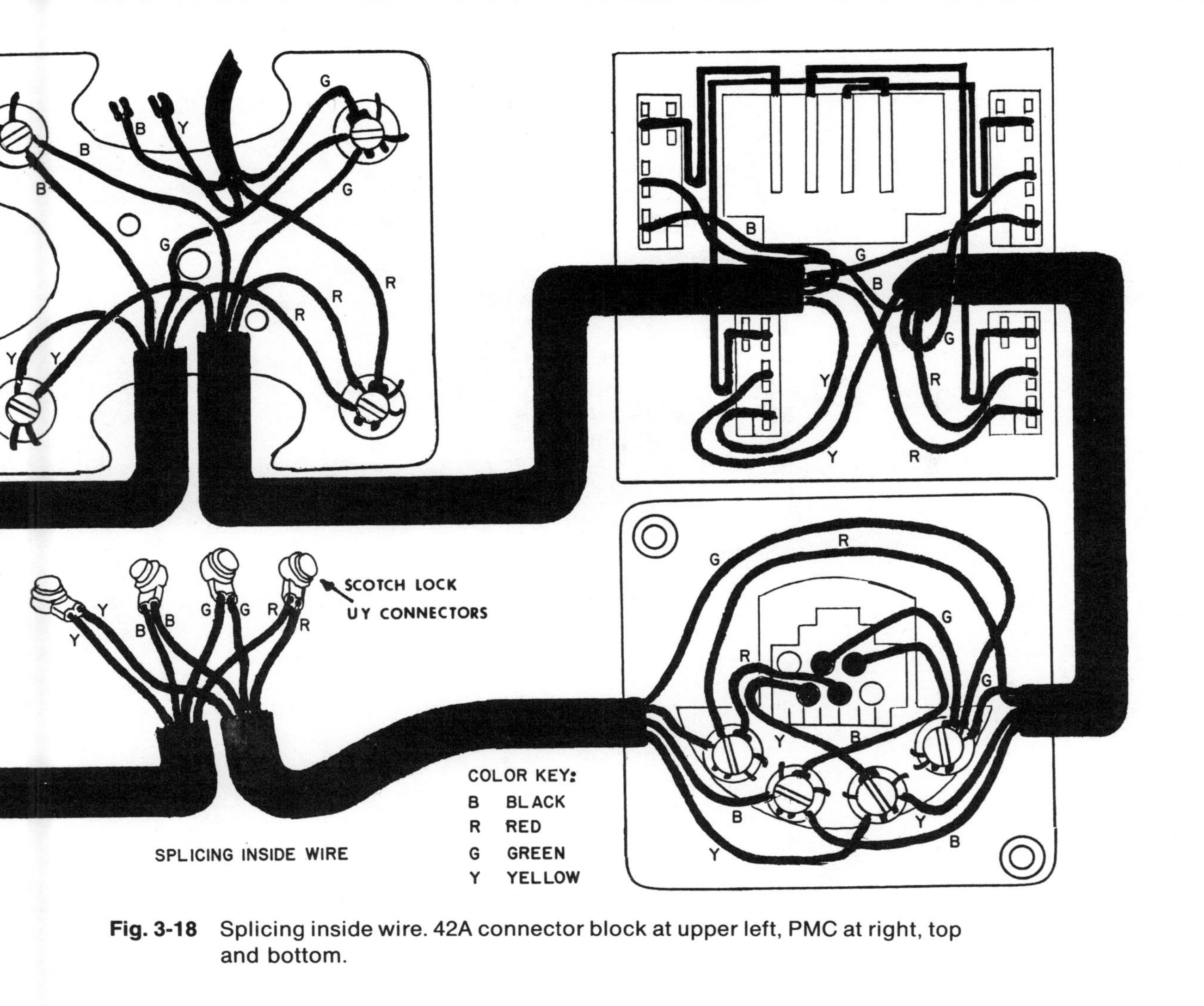

Fig. 3-18 Splicing inside wire. 42A connector block at upper left, PMC at right, top and bottom.

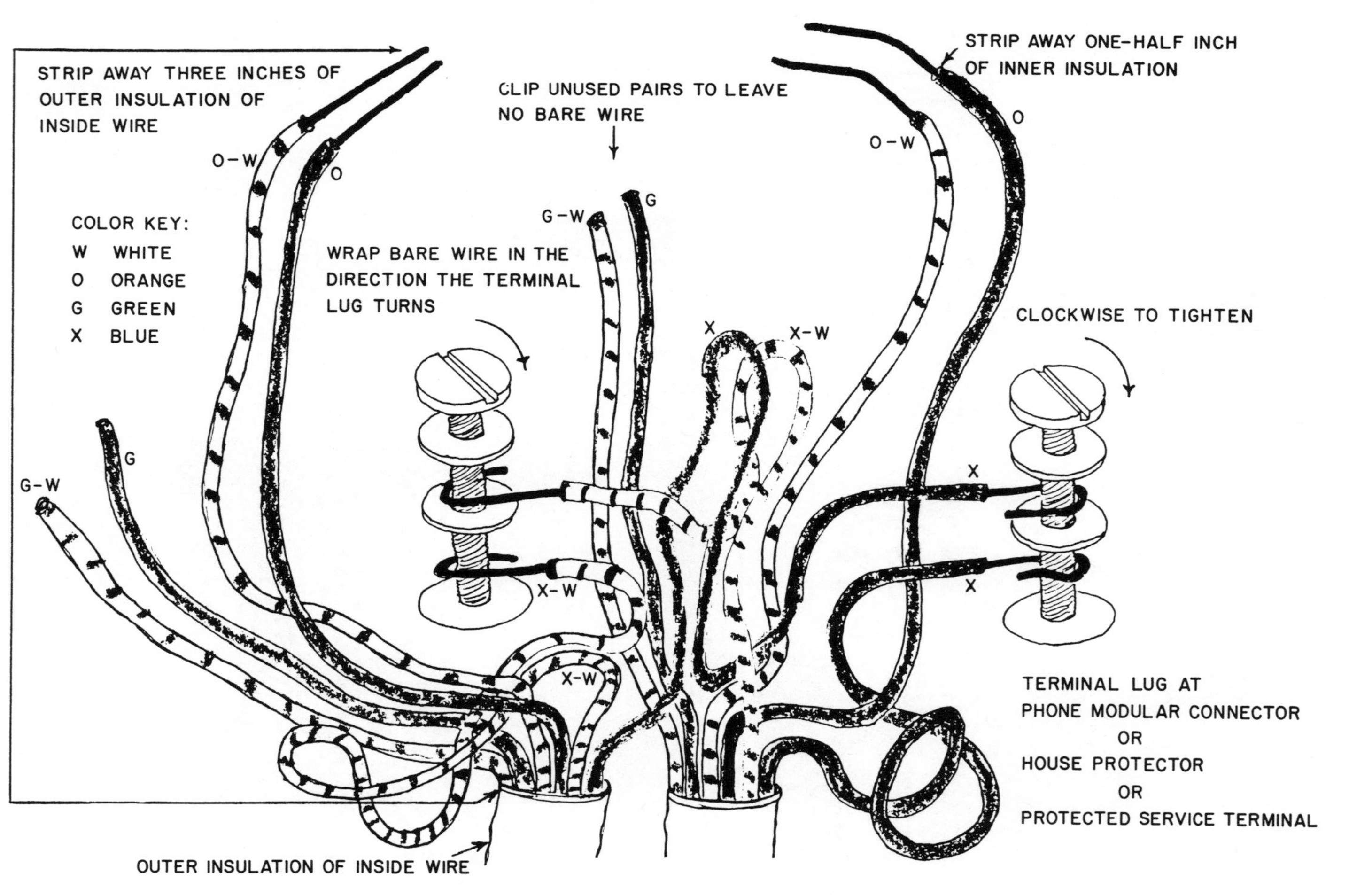

Fig. 3-19 Splicing inside wire to terminal lugs.

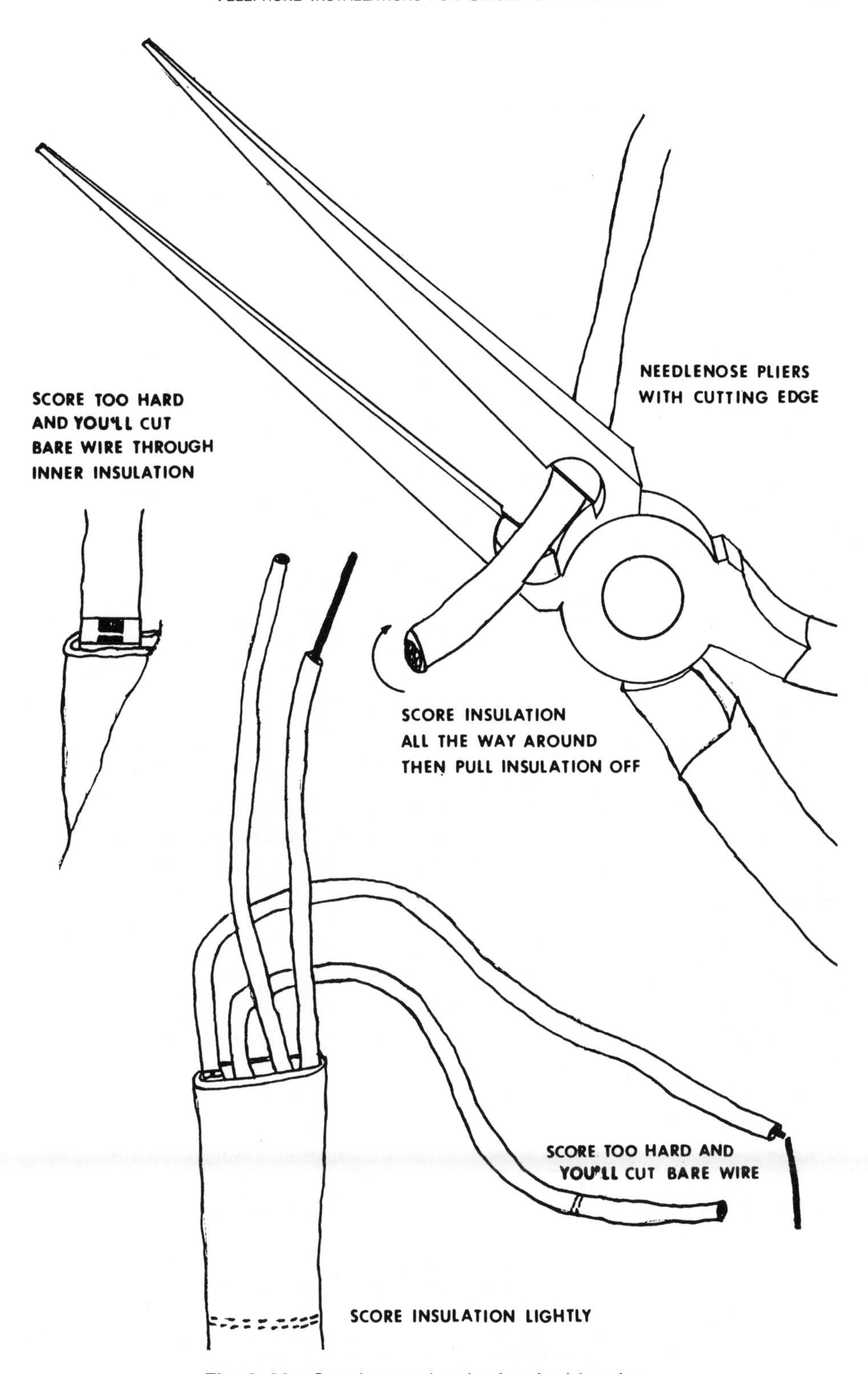

Fig. 3-20 Scoring and stripping inside wire.

outer insulation layer all the way around, lightly bite into this layer and pull it off. (By scoring too hard, you may cut the bare wire inside the inner insulation layer.) For the inner insulation, it is not necessary to score all the way around; just lightly bite into the insulation and pull it off. Stripping inside wire does take practice, so try it a few times on a small piece of wire before making the final connection.

Always clip unused pairs (see Figure 3-19); otherwise, the bare wires may make contact with the working pair and short or ground it out. It is also best to leave unused pairs the same length as working pairs in case you need to use them at a later date.

Standard Color Code for Connecting Inside Wire to the Phone Modular Connector

The phone modular connector provides a national telephone standard for phone jacks, allowing telephones to be attached and detached at will. Part of that national standard are the pin designation and lead colors for the phone modular connector.

By closely examining the phone modular connector, you will see either four or six copper pins in the slot where you plug in your telephone. Figure 3-21 shows a phone modular connector with six pins, labeled 1 through 6, and the accompanying color code for each pin connection; Figure 3-22 shows four pin connections, labeled 2 through 5. The pins labeled 3 and 4 in both figures are the standard connecting points for attaching inside wire—these two pins are the two points where you connect the pair for a standard connection.

Pins 3 and 4 have the corresponding color codes of red and green. Take the pair you connected at the house protector and connect it to pins 3 and 4 in your phone modular connector. To keep continuity with the color code, use the red and green wires of the inside wire for your pair.

Once the pair is connected and a phone plugged in to the phone modular connector, you should hear the dial tone and be able to call out. If there is no dial tone, either you connected the wrong pair from the house protector to the phone modular connector or your house protector may not be activated, requiring the services of the telephone company. If all systems are "go" and you still don't have a dial tone, check the inside wire for a break.

If you hear a dial tone but are unable to dial out, change the position of the pair at either the house protector or at the phone modular connector pins. Some telephones require a certain polarity in order to dial out; by switching the wires around, proper polarity is provided.

Variations of the Standard Color Code

There are a few variations from the standard connection for inside wire to the phone modular connector that can eliminate the need to install additional inside wire and phone modular connectors.

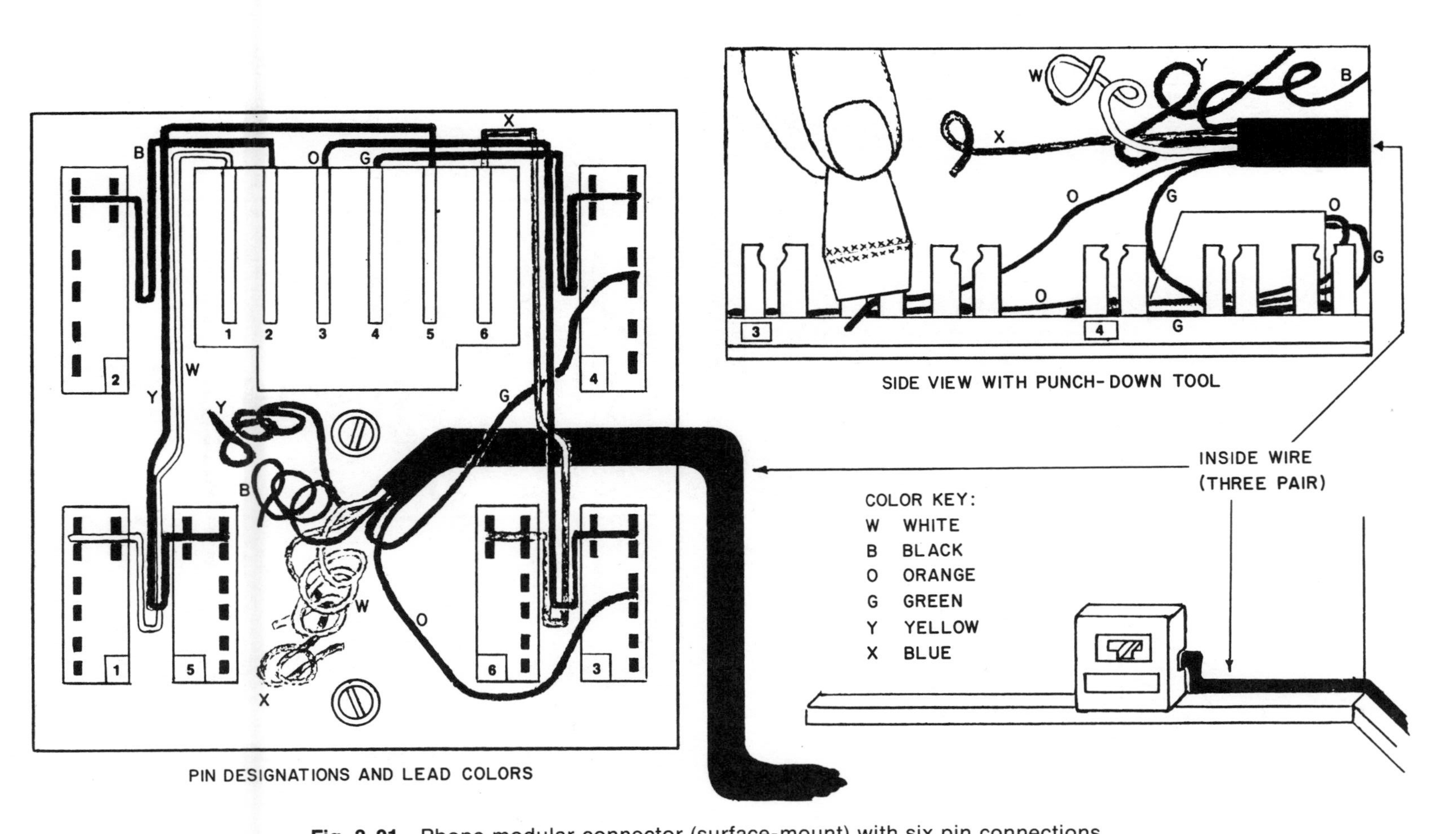

Fig. 3-21 Phone modular connector (surface-mount) with six pin connections.

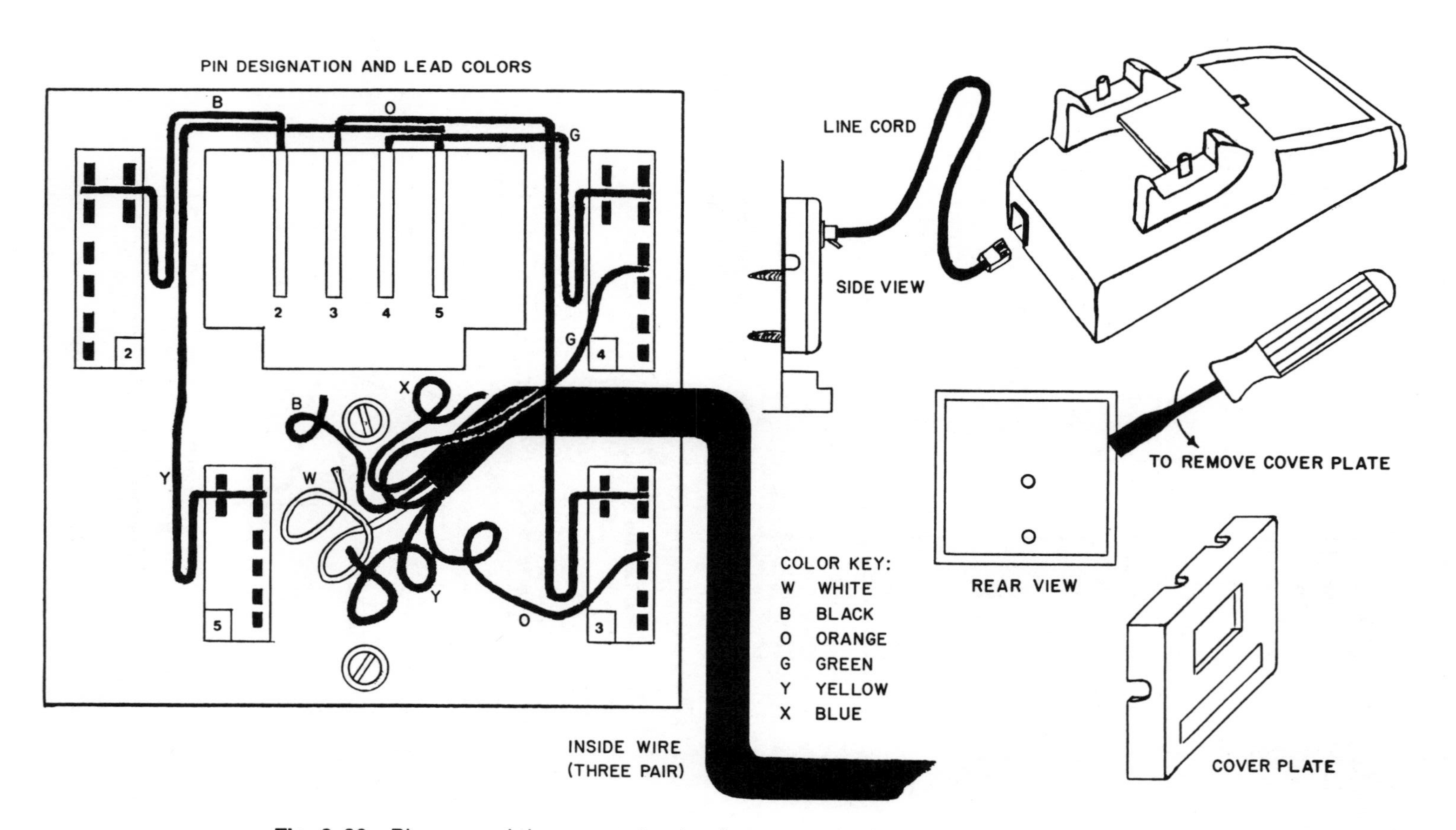

Fig. 3-22 Phone modular connector (surface-mount) with four pin connections.

Using Only One Phone Modular Connector for Two Telephone Numbers

For one phone modular connector to be used for two telephone numbers, the following conditions must be met:

1. The existing phone modular connector location must be suitable for the second telephone number.
2. The existing inside wire at the phone modular connector must contain an extra pair that travels back to the house protector or protected service terminal. (To determine this, see Chapter Five, "Test Equipment" and Chapter Four, "Protected Service Terminal.")
3. The telephone instrument you intend to use for the additional number must have a four-wire line cord, and entry into the instrument must be possible. If the line cord and entry requirement are not met, you must have a telephone instrument capable of accepting two independent telephone numbers.
4. If your existing telephone instrument has a lighted dial or touch pad, which requires an external power source, your line cord must contain six leads and your phone modular connector must contain six pin positions (see Figure 3-21). However, there are a few telephone instruments that use the voltage existing on the pair to light dials or touch pads, in which case, you should have no problem.

The standard color code connections at the phone modular connector are pin positions 3-red and 4-green (see Figures 3-21, 3-22 and 3-23). By attaching the second pair for the additional telephone number to pin positions 2-black and 5-yellow, you have just wired the phone modular connector to accept two independent telephone numbers. (Figure 3-24, a variation of the standard color code connection, details how the phone modular connector looks when wired to accept two independent telephone numbers.)

The next step is to wire the telephone instrument to accept these new pin positions. If your telephone is already capable of accepting two telephone numbers independently, this step is not required. If not, complete the following instructions.

In order to wire the telephone instrument to accept pin positions 2-black and 5-yellow, you must open the telephone instrument and follow the line cord into the network, where you will find two wires, one green (pin position 4) and one red (pin position 3), connected to the network. You should also see two extra wires, one black (pin position 2) and one yellow (pin position 5), that are not connected but extend from the line cord. Exchange the red wire (pin position 3) with the yellow wire (pin position 5), and then exchange the green wire (pin position 4) with the black wire (pin position 2). You have just completed the steps required to wire your telephone instrument to accept the second telephone number at the phone modular connector.

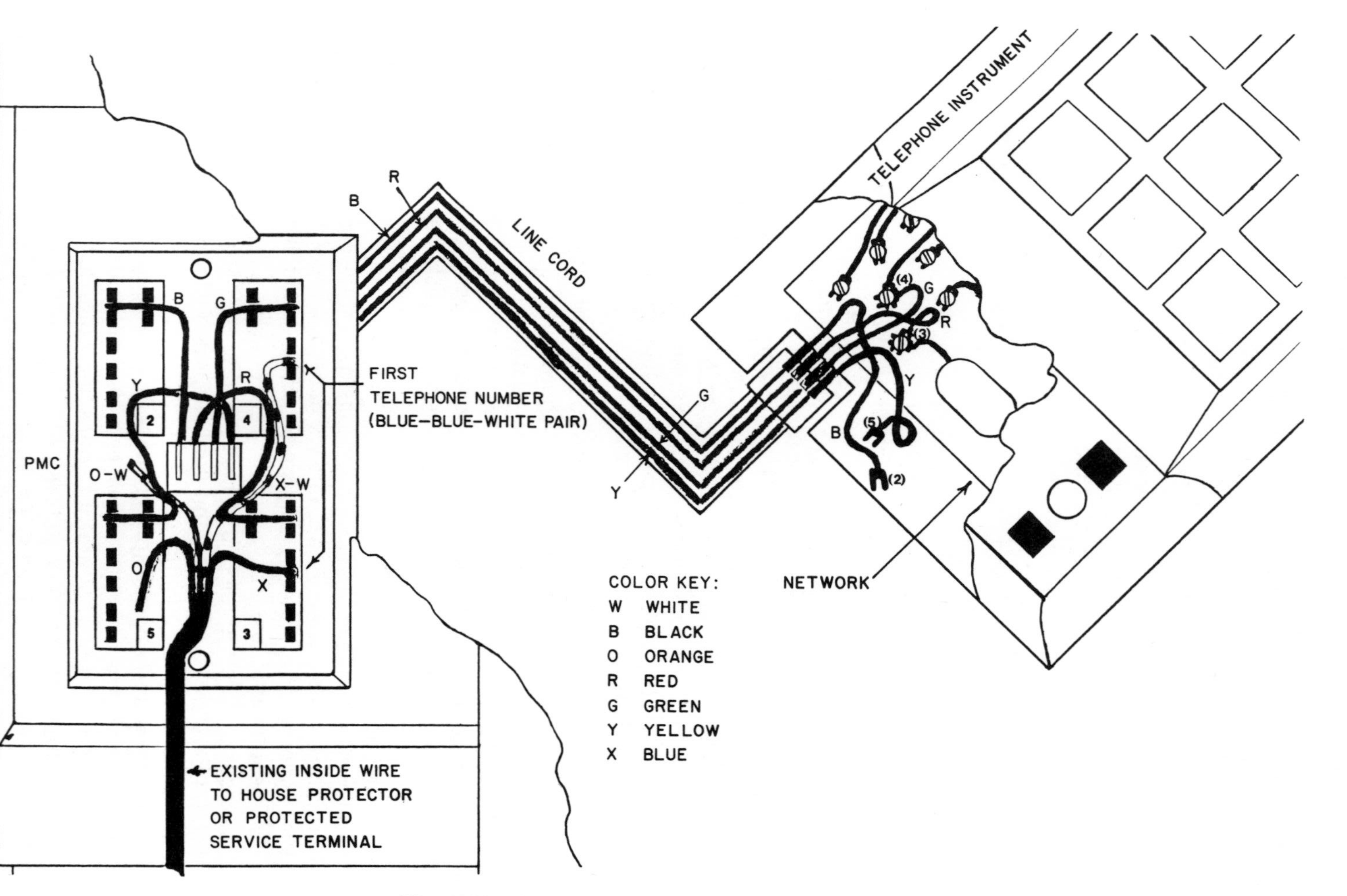

Fig. 3-23 Standard color code connection.

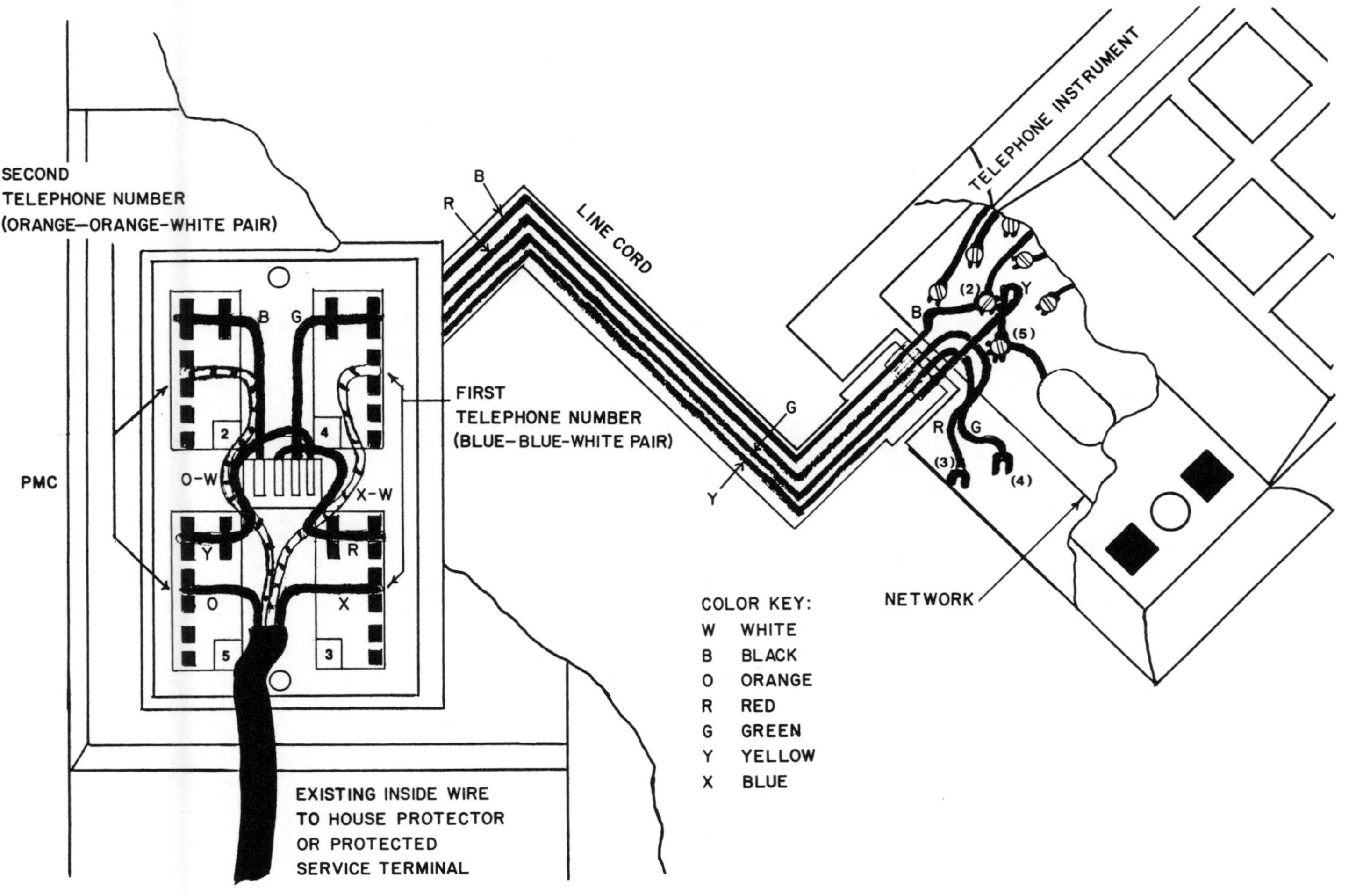

Fig. 3-24 Variations of the standard color code connection.

Before closing the telephone instrument, tape the green and red wire leads so they will not make contact with one another and short the first telephone number.

With the inside wire, phone modular connector, and telephone instrument now wired to accept an additional telephone number, consideration must be given to the house protector or protected service terminal. For single-unit dwellings where service is provided by general telephone cables, adding a second telephone number will require an additional drop wire and house protector. Figures 3-25 and 3-26 detail how a second telephone number may be connected to your inside wiring with two single-station or one multistation house protectors.

Two or More Telephone Numbers

You may find that wiring a phone modular connector to accept two telephone numbers does not meet your needs for various reasons: existing phone modular connector locations are not suitable; your telephone instrument cannot be entered to allow for the necessary line cord changes; or you require two or more telephone numbers. If any of these are the case, the following techniques offer alternative solutions designed for the least amount of work at the least expense.

Figure 3-27 details the required phone modular connector connections needed to provide three independent telephone numbers, all of which connect to the standard 3- and 4-pin positions. This standard connection does not require you to change the line cord leads but does require that the second and third pairs (assuming that the first pair is a working number) travel from the phone modular connector back to the house protector or protected service terminal. (See Chapter Five, "Test Equipment" and Chapter Four, "Protected Service Terminal.") Although Figure 3-27 shows all three phone modular connectors placed next to one other, this does not necessarily have to be the case (see Figure 3-12). With an in-series hookup, it is possible to extend inside wire from an existing phone modular connector to provide an extension to any location you desire. After splicing inside wire, your only limitation in extending a phone modular connector location will depend entirely upon the surface and structural conditions existing in your house.

Another technique that combines two previously explained methods is illustrated in Figure 3-28, which details connecting phone modular connectors to accept two independent telephone numbers and to be wired for the standard connections of pins 3 and 4. Note in Figure 3-28 that in order to utilize the third telephone number at two of the phone modular connector lines, cord changes must be made at the telephone instrument unless you have an instrument capable of accepting two numbers.

Connecting Lighted Dials and Touch Pads

When one telephone number is involved, a connection for a phone with a lighted dial or touch pad is standard, requiring no changes at the line cord (Figure 3-29). Note that pin positions 2-black and 5-yellow provide the necessary electrical current to light the dial or touch pad. Since

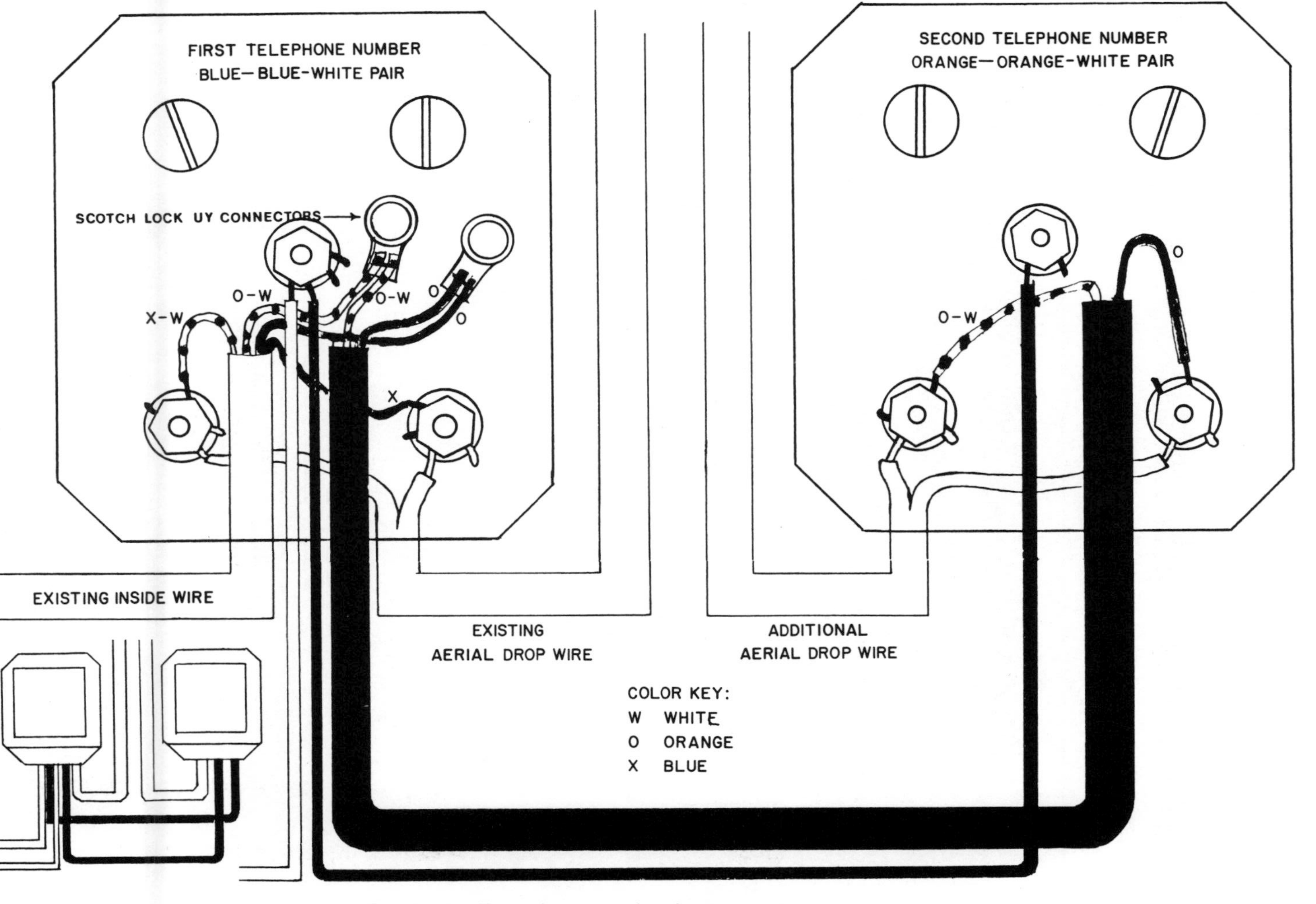

Fig. 3-25 Two single station house protectors.

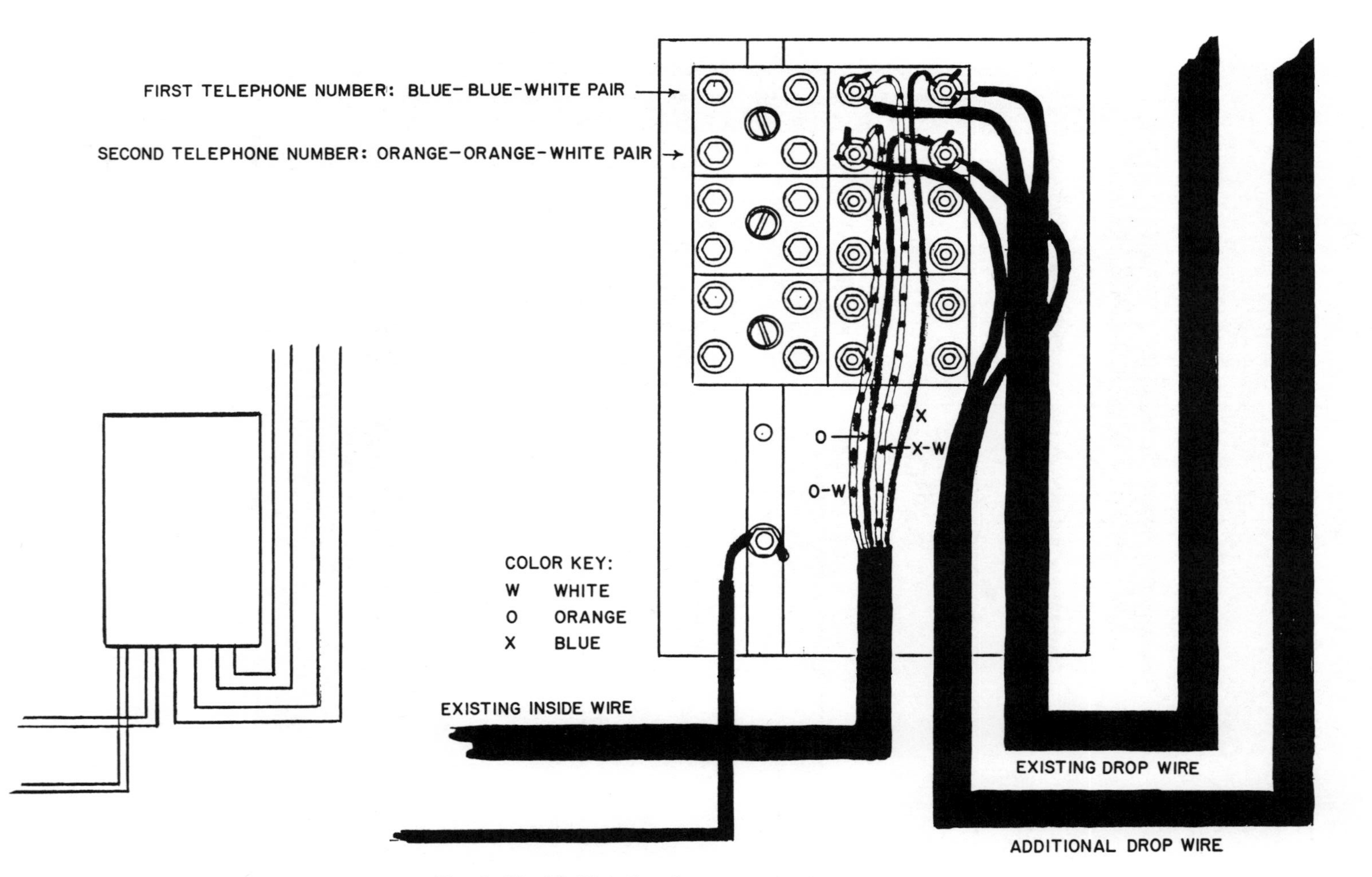

Fig. 3-26 Multistation house protector.

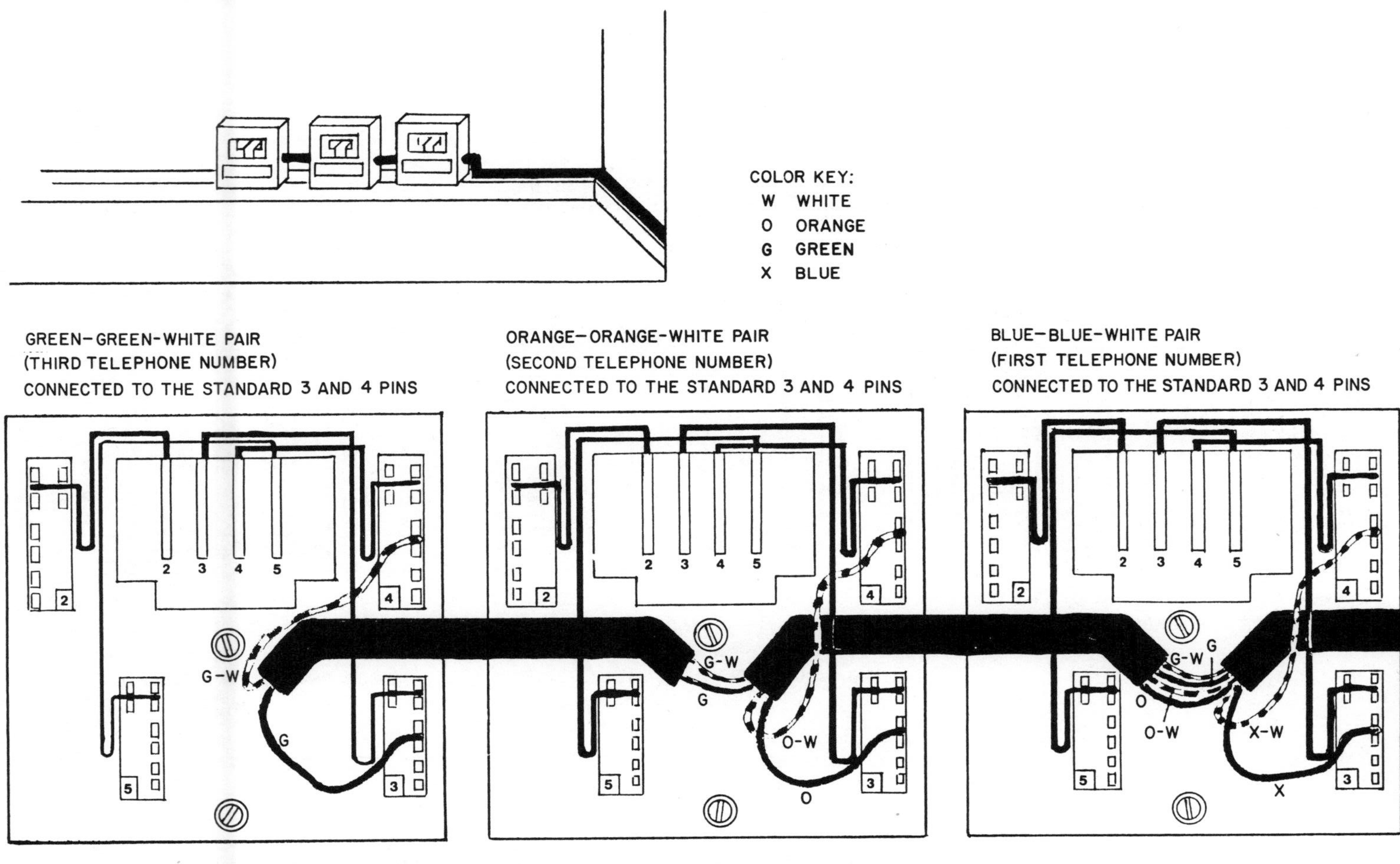

Fig. 3-27 Variations of the standard color code connection—one arrangement using existing three pair inside wire to provide three independent telephone numbers.

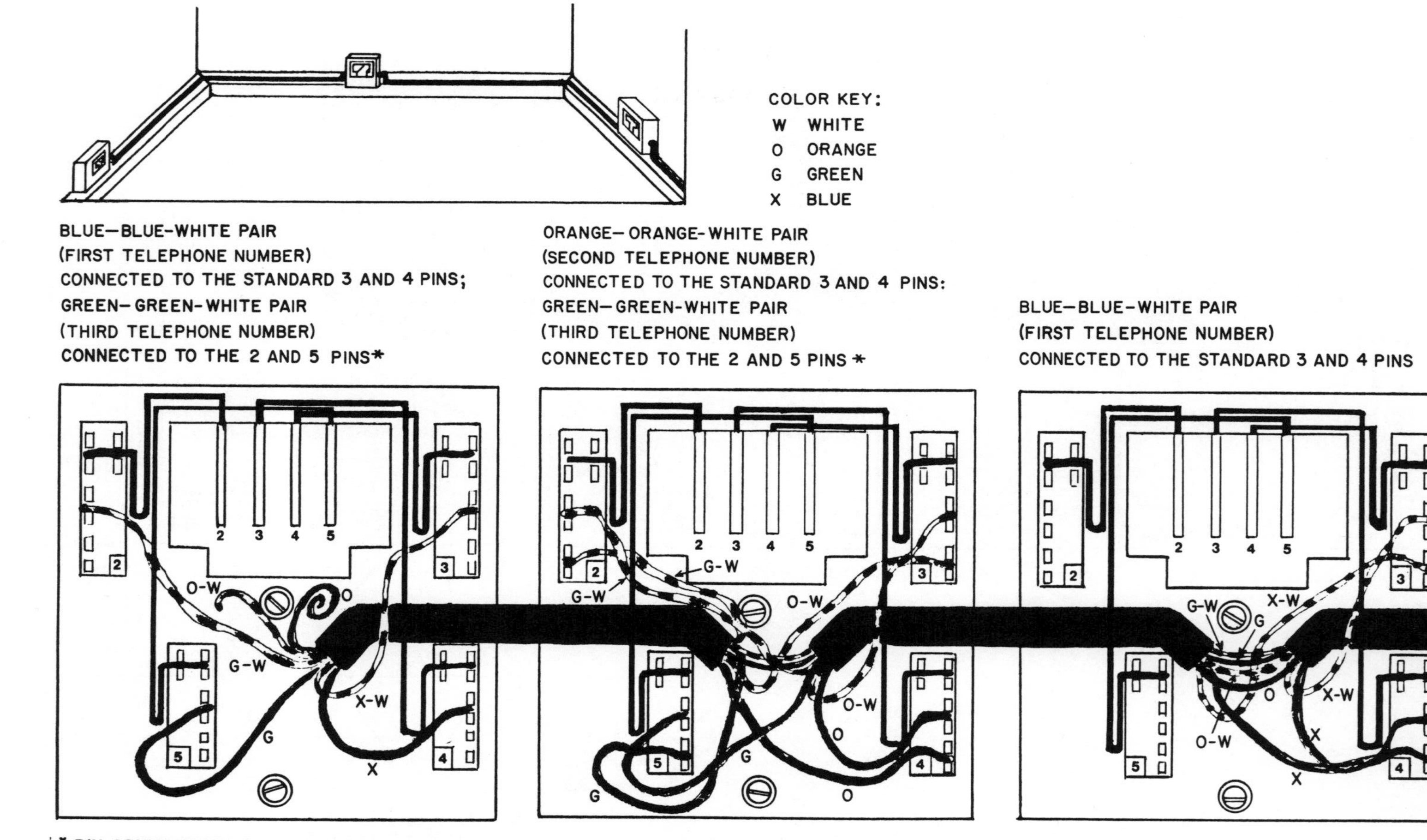

Fig. 3-28 Variations of the standard color code connection—another arrangement using existing three pair inside wire to provide three independent telephone numbers.

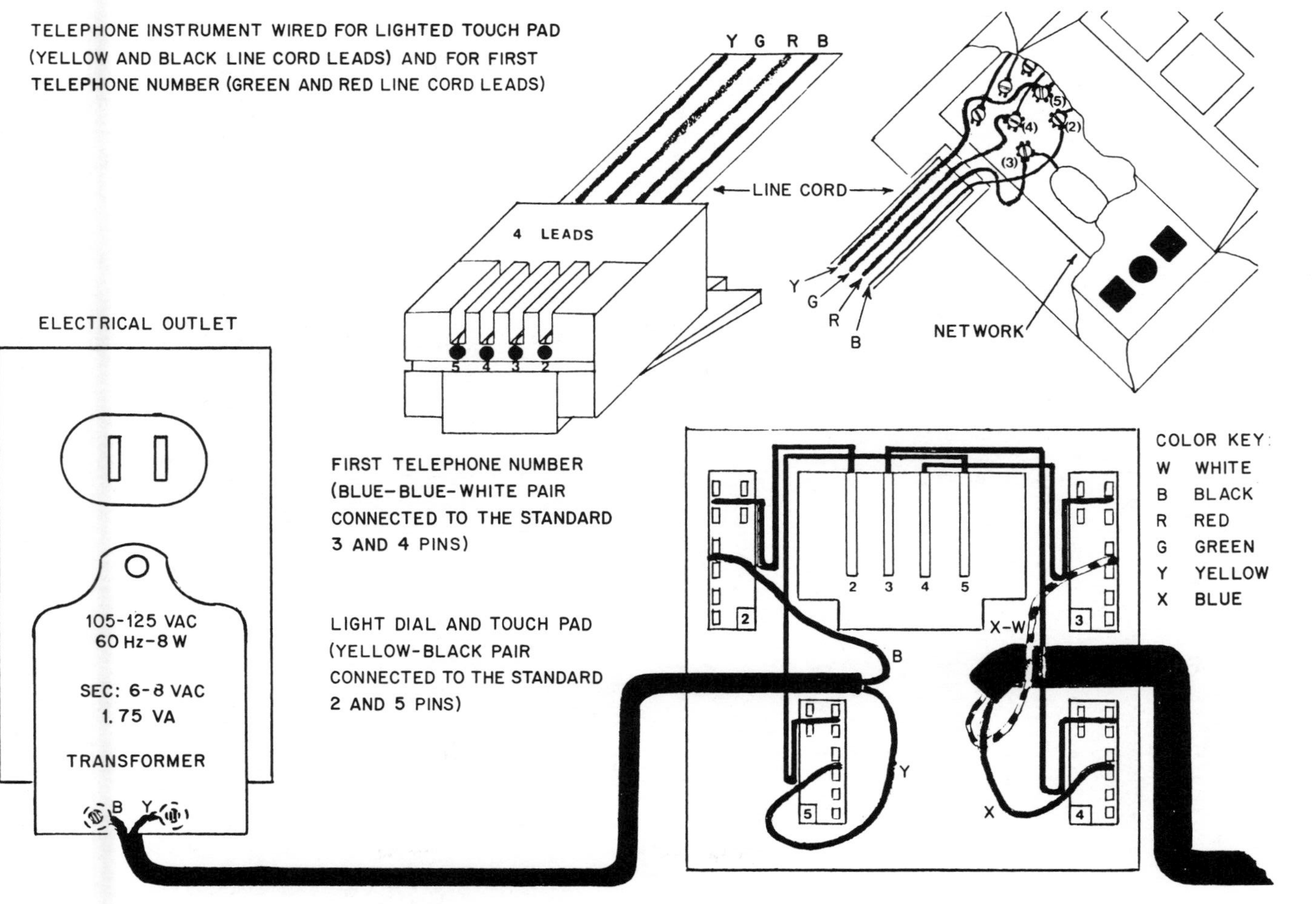

Fig. 3-29 Lighted dial and touch pad connection.

the same two pins are needed to accommodate a second telephone number, it is not possible to have a lighted touch pad or dial along with a second number at the same phone modular connector. However, if you have a 6-pin phone modular connector and a 6-lead line cord connecting a second telephone number, having a lighted dial or touch pad would be possible (Figure 3-30). Pin positions 2-black and 5-yellow are the connecting pins for the power supply. To wire the phone modular connector to utilize a second telephone number, pin positions 6-blue and 1-white (see Figure 3-21) need to be used and line cord leads changed, as shown in Figure 3-30. Exchange the green lead (pin position 4) with the white lead (pin position 1), and exchange the red lead (pin position 3) with the blue lead (pin position 6). Now the telephone instrument will have power for the lighted dial or touch pad and also be able to accept the second telephone number.

Different Types of Phone Modular Connectors

Regardless of the size, shape, or purpose of a phone modular connector, they all have the same pin designation, lead colors, and modular line cord receptacle size. One way to better understand the different types, though, is to review phone modular connectors according to their design and purpose.

The flush-mounted phone modular connector, designed to be attached to an existing wall fixture and to fit flush against the wall, is for new installations (using the prewire method); for the replacement of existing flush-mounted hard-wired phone jacks; or for the placement of wall-mounted telephones. Refer to Figure 3-9, which shows that flush-mounted phone modular connectors need wall-mounted fixtures to be placed. It details the add-on wall fixture, which changes a flush-mounted phone modular connector to a surface-mounted one. With this change, you will have the flexibility to attach a wall-mounted modularized telephone anywhere an inside wire can reach. The locking pins are required to hold the wall-mounted telephone tight to the phone modular connector.

Without the add-on wall fixture, you will need to have built-in type wall fixtures to which to attach the flush-mounted phone modular connector. The same as electrical outlet boxes, the built-in type wall fixtures are available at most electrical supply stores.

Looking back at Figures 3-23 and 3-24, you'll see the rear view of a flush-mounted phone modular connector mounted in the wall (the wall fixture is omitted for clarity). The type shown is a punch-down phone modular connector, with which there is no need to strip the inner insulation on the pairs. A punch-down tool is provided with the purchase of this type of connector (see Figure 3-21).

The flush-mounted duplex phone modular connector (Figure 3-31) provides the capability to accept four independent telephone numbers. (To wire it, refer back to "Using Only One Phone Modular Connector for Two Telephone Numbers" in this chapter.) This type of phone modular connector can be wired with the same telephone number at both recepta-

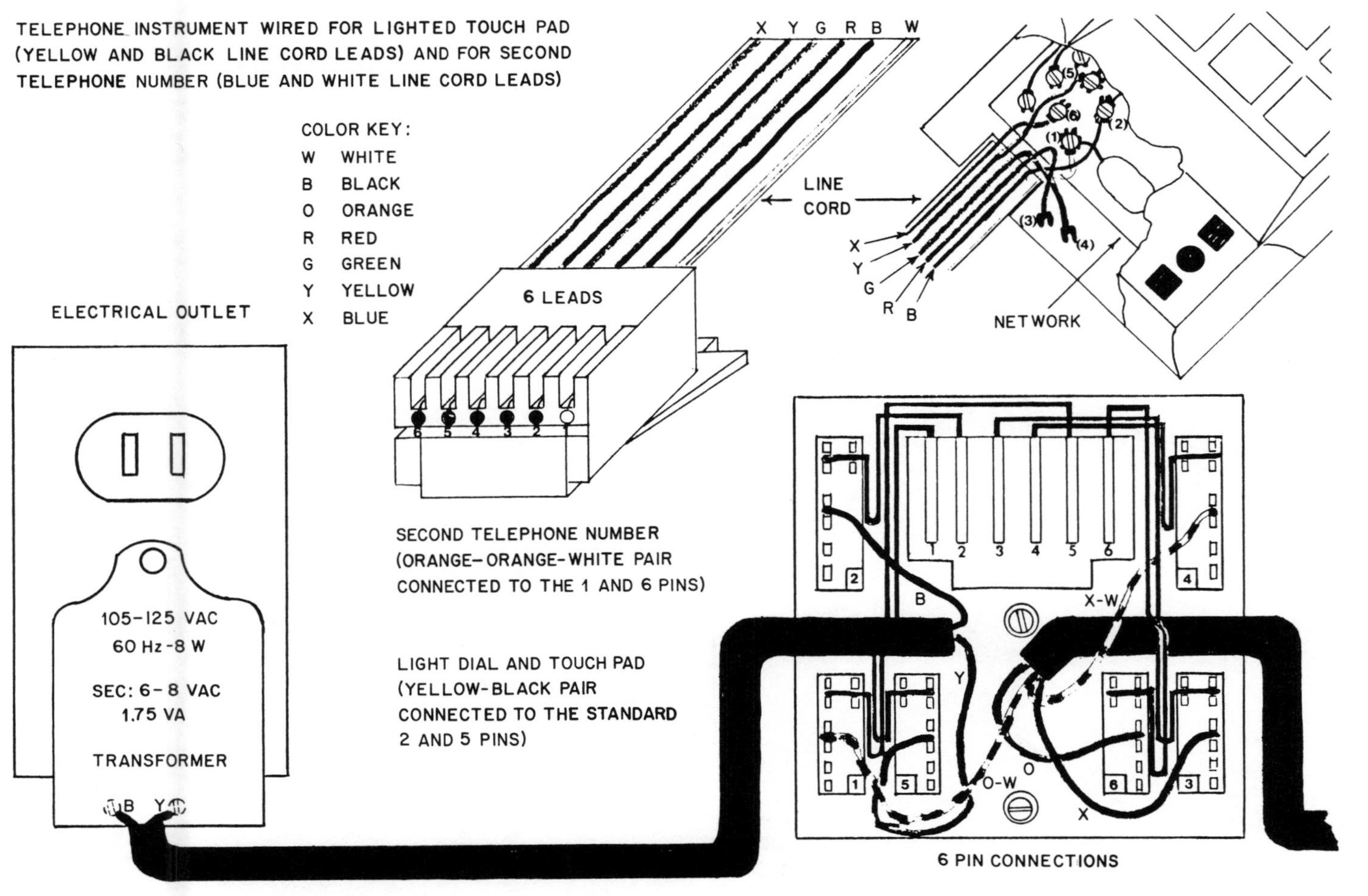

Fig. 3-30 Lighted dial and touch pad connection with second telephone number.

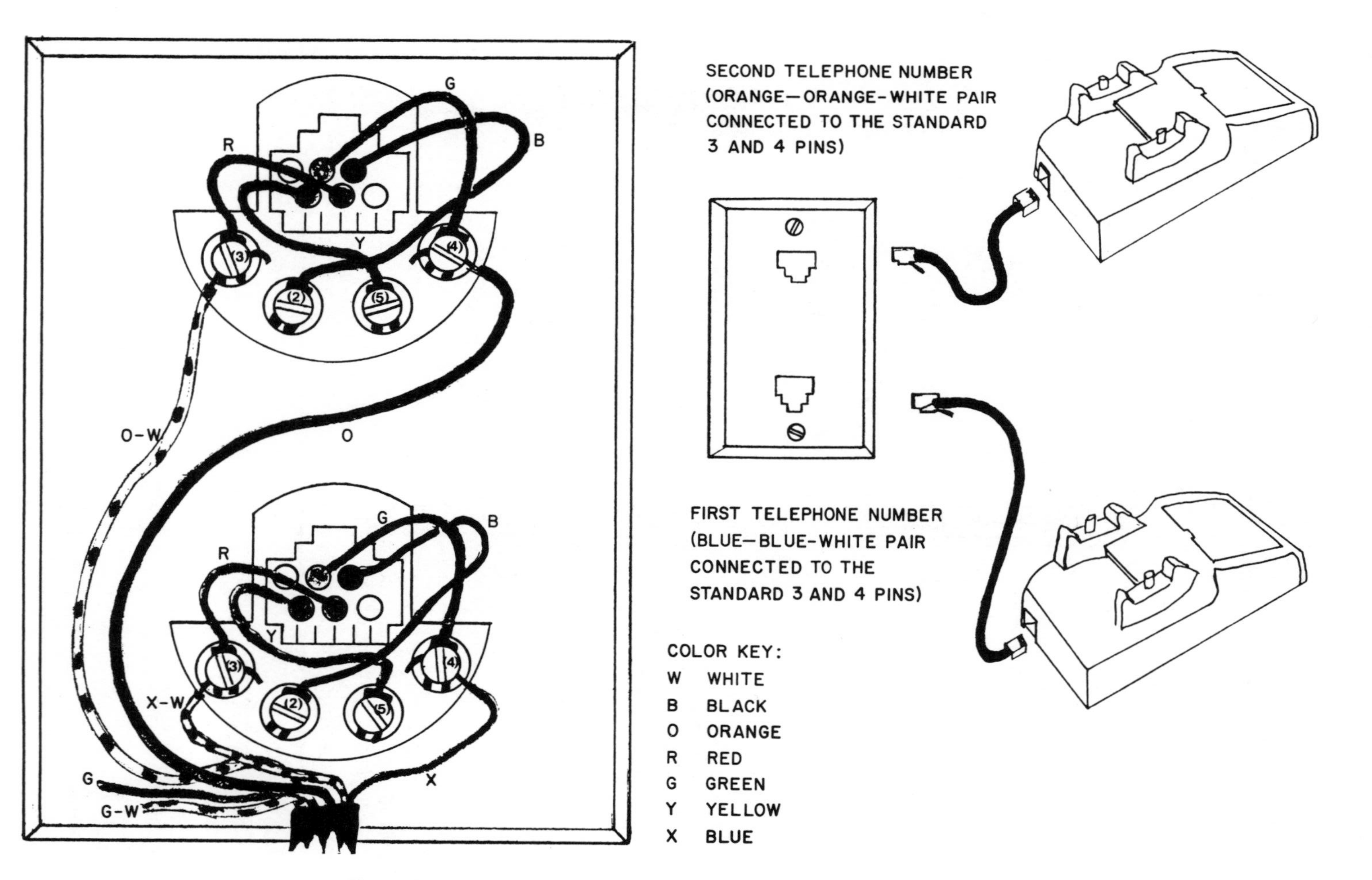

Fig. 3-31 Duplex phone modular connector—flush-mount.

cles, allowing a telephone and an answering machine to work together. It can also be wired to accept two independent telephone numbers with the standard connections at pins 3 and 4; thus, there will be no need to make line cord lead changes (see Figure 3-31).

A flush-mounted phone modular connector that doubles as a surface-mounted type (Figure 3-32) provides its own numbering system for wiring instructions; however, the lead colors are the same as on all other phone modular connectors. The illustration also includes the numbered pin positions corresponding to their lead wires. The cover plate comes with or without locking pins for wall-mounted telephones. This type of connector is a punch-down type where no wire stripping is necessary.

The surface-mounted phone modular connector is designed for new installations (utilizing the wrap method or the run-under method) and for the replacement of existing surface-mounted hard-wire phone jacks (Figure 3-33). This type is designed to be placed near the floor at the baseboard. (The one shown in the illustration requires wire stripping because it has terminal lugs.)

Another surface-mounted phone modular connector (see Figures 3-21 and 3-22) is a punch-down type where no wire stripping is necessary. The backplate, which has all the wiring, can be attached in place while the cover plate is removed, allowing you to run the inside wire to the phone modular connector in order to connect your pair and test it. This is the easiest method since the other type has to be removed from the wall for connecting or testing.

Frequently, removing a connector leaves a loose jack, which will eventually fall off the wall. With the punch-down type, once the base plate is secure, the cover plate can be removed many times to allow access to the wiring without jeopardizing its tight and secure placement. Not as thick as others, this type of phone modular connector is less likely to be hit by a vacuum cleaner moving along the baseboard (vacuums are responsible for damaging a great many surface-mounted connectors).

The surface-mounted conversion phone modular connector (Figure 3-34) is designed to replace existing surface-mounted hard-wired phone jacks (see Figure 3-37). When installing this type of connector, simply remove the hard-wired phone jack cover, remove the hard-wired line cord, and then connect the color coded leads from the conversion phone modular connector to the terminal lugs on the remaining 42A connector block.

When a duplex phone modular connector adapter (Figure 3-35) is installed, it actually provides another phone modular connector. The pin designation points are doubled, enabling you to wire one phone modular connector for two telephone numbers (see Figure 3-24). This adapter also provides the answer for having an answering machine and a telephone working at the same location.

One of the first attempts by the telephone company to modularize telephone instruments resulted in the four-prong phone plug (Figure 3-36). Although it is technically a phone modular connector, it is more commonly known as a phone plug. Illustrated are the adapters available to

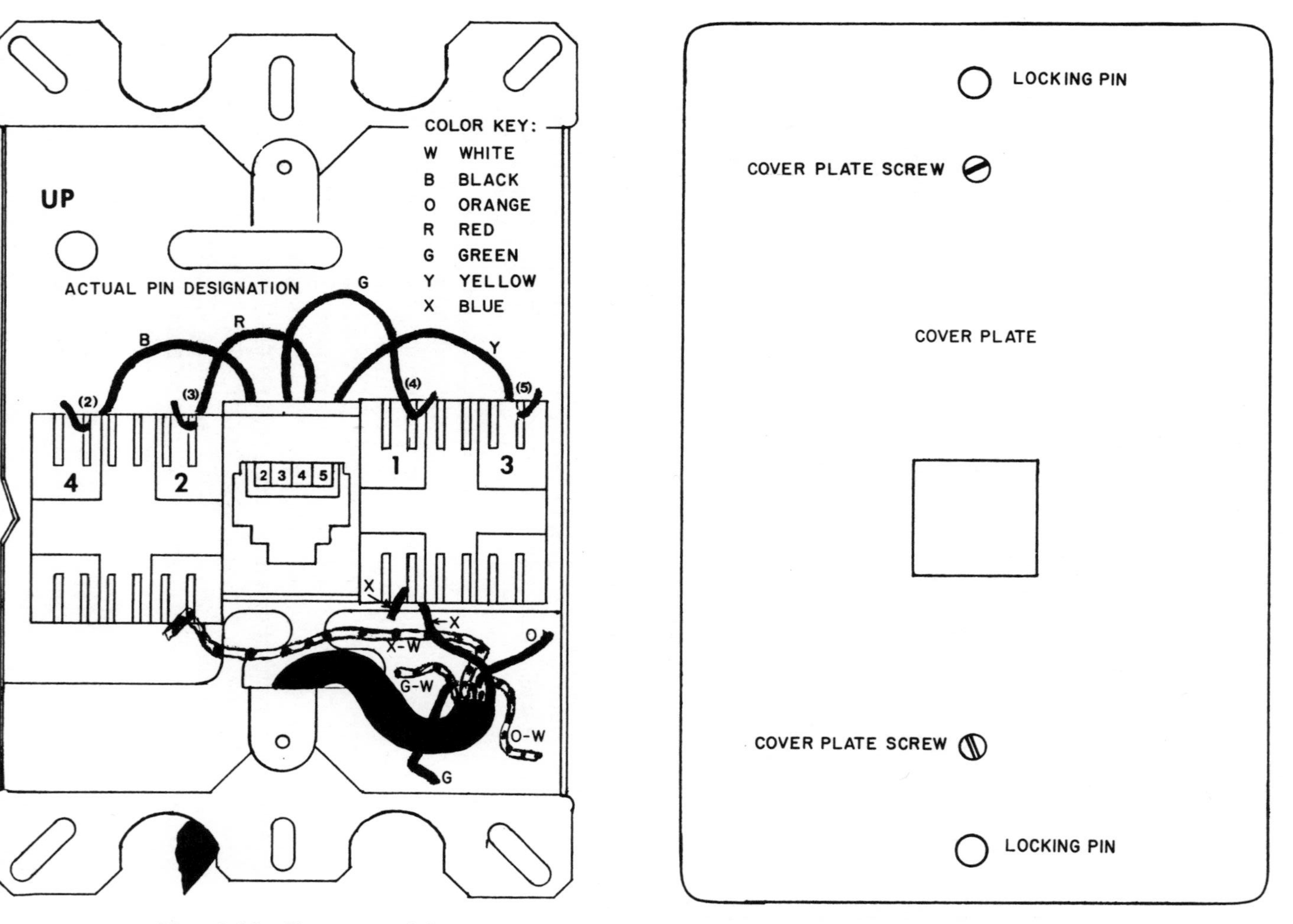

Fig. 3-32 Phone modular connector—surface-mount and flush-mount—designed for wall-mounted telephone instruments.

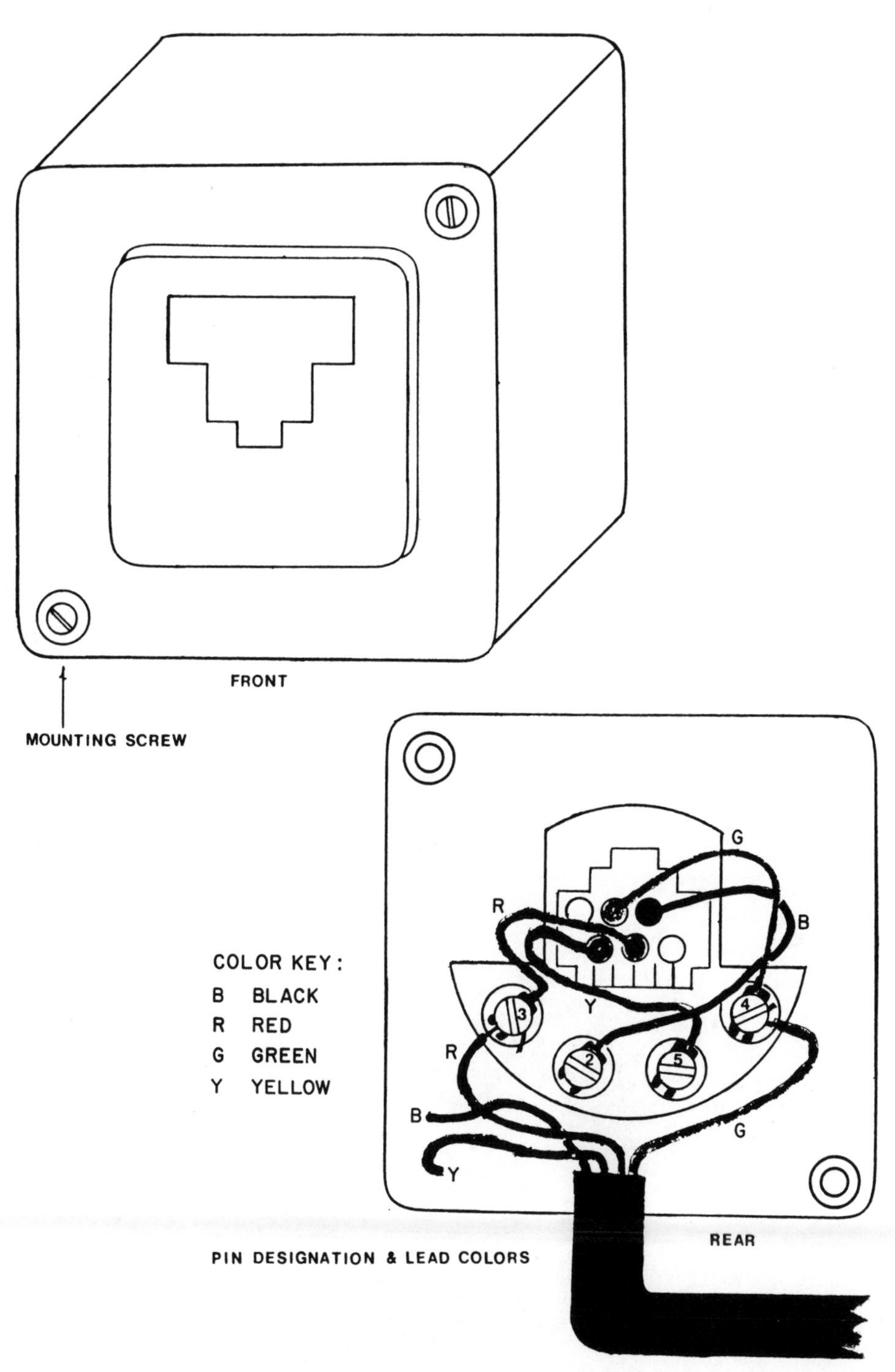

Fig. 3-33 Phone modular connector—surface-mount—designed for replacement or installation.

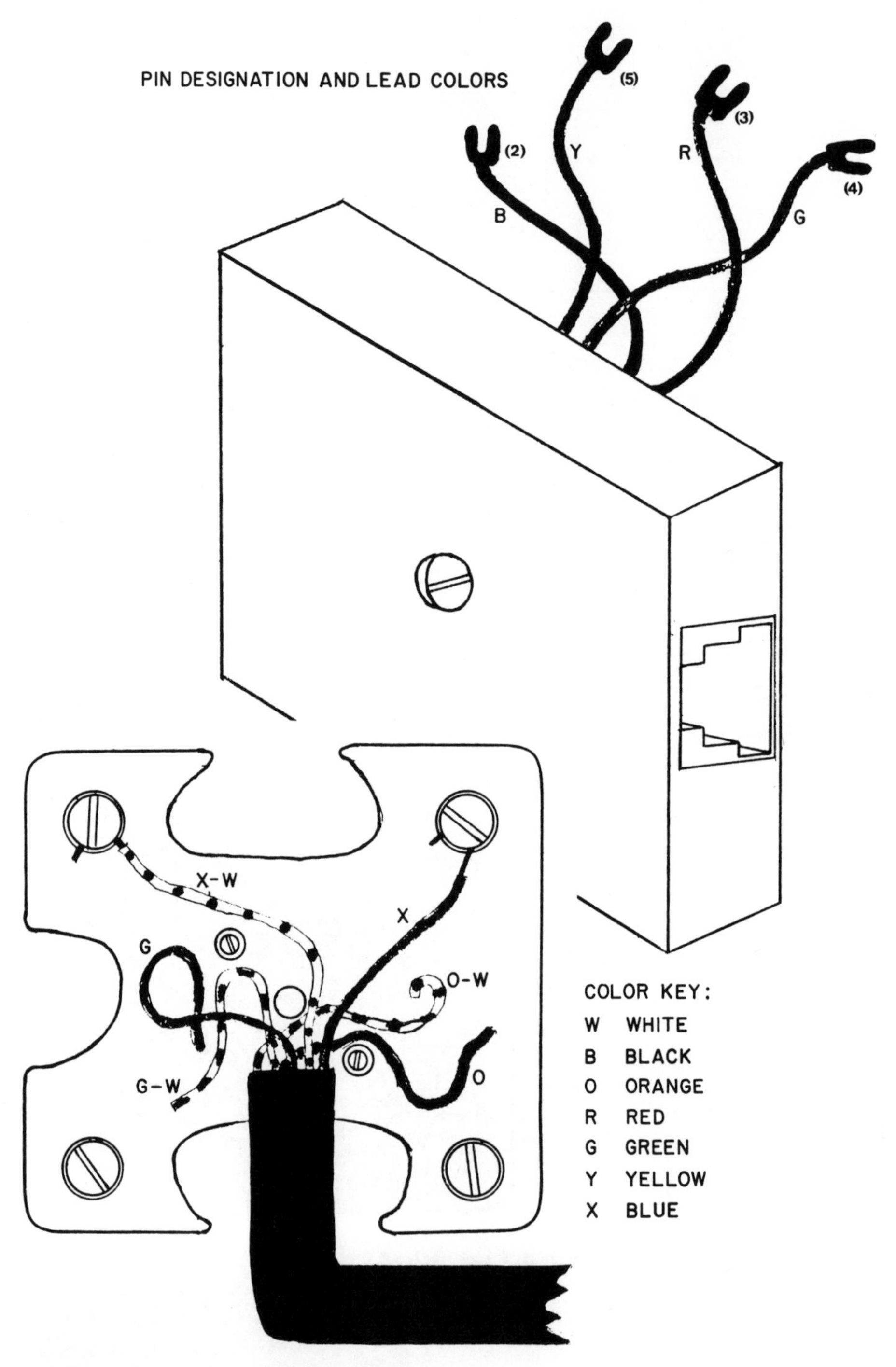

Fig. 3-34 Phone modular connector—surface mount—designed for conversions.

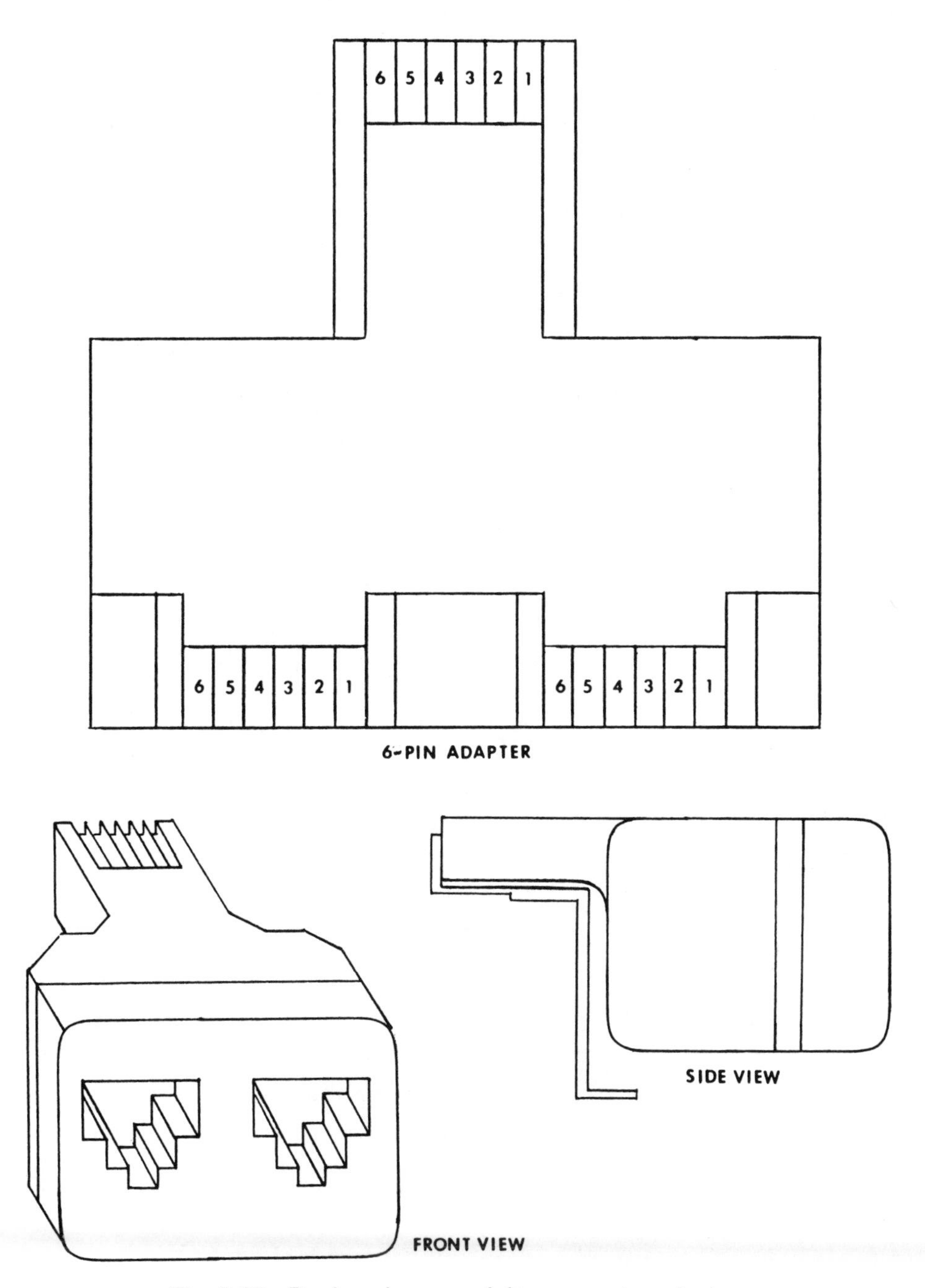

Fig. 3-35 Duplex phone modular connector adapter.

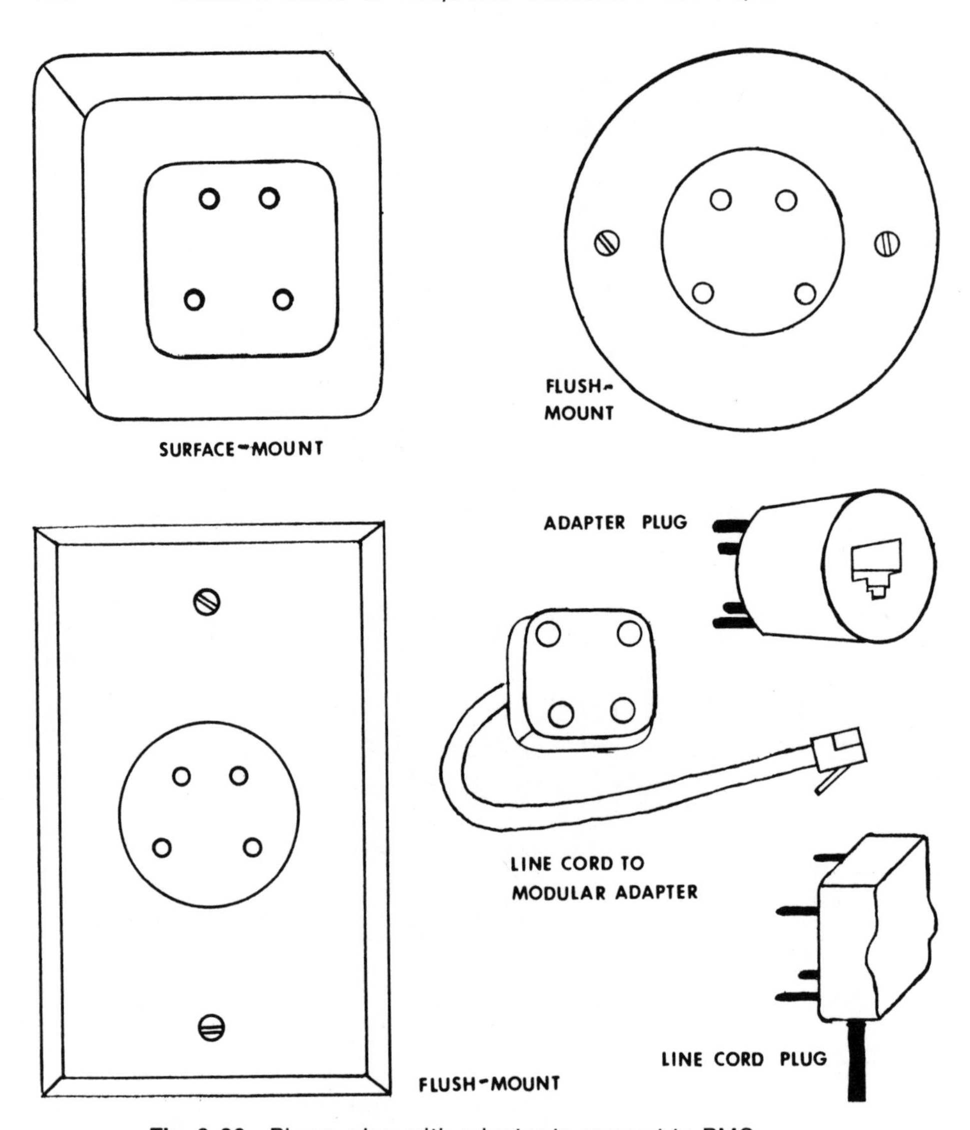

Fig. 3-36 Phone plug with adapter to convert to PMC.

convert four-prong phone plugs to the national standard phone modular connector.

Hard-Wire Phone Jacks

Before the modularization of phone jacks and telephone instruments, the telephone company connected instruments to the inside wire by means of a permanent phone jack, commonly referred to as a hard-wire phone jack (Figures 3-37 and 3-38).

With the hard-wire phone jack the telephone instrument cannot be easily disconnected from the telephone circuit for isolating trouble or repair and cannot be moved to another location, thereby losing mobility.

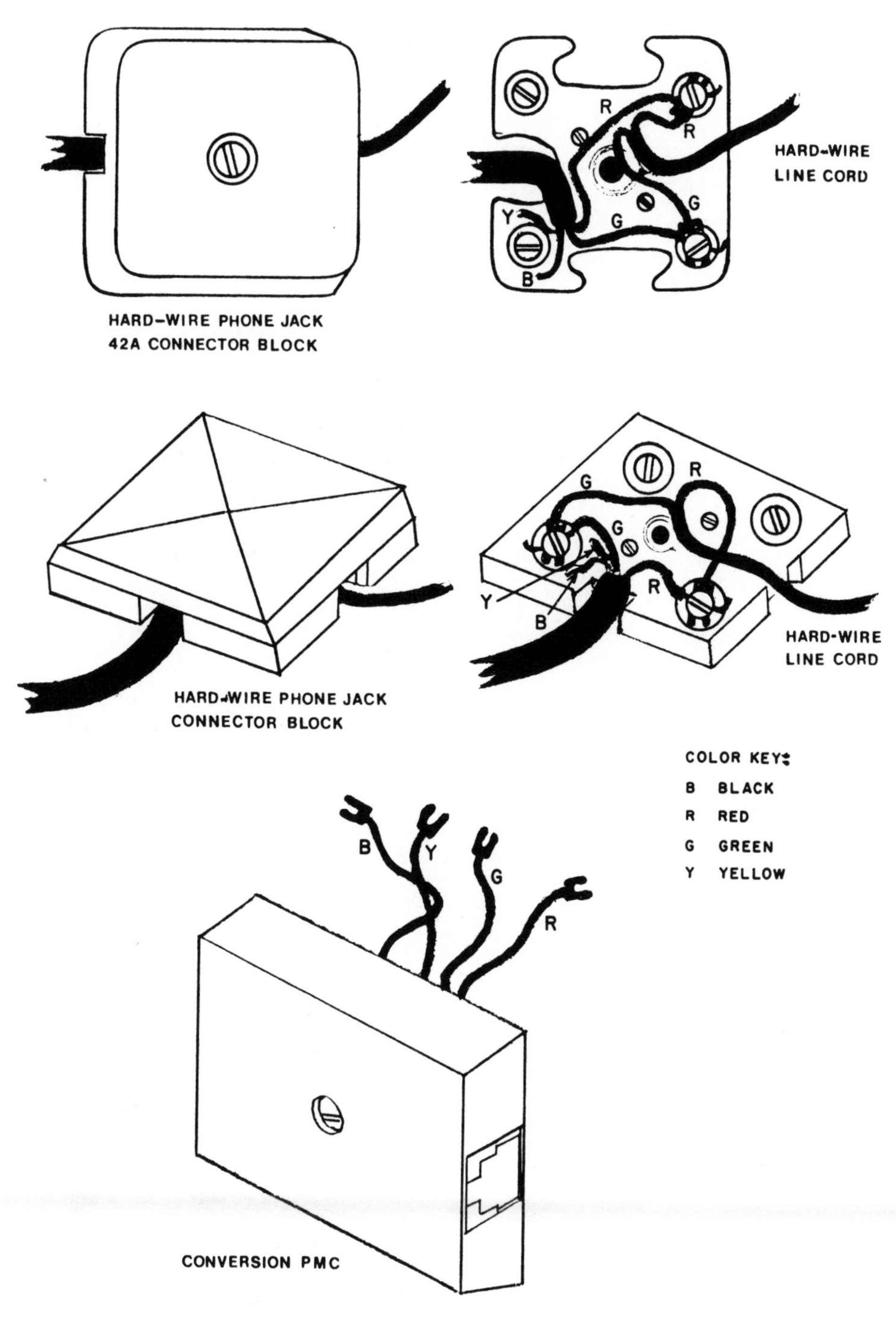

Fig. 3-37 Hard-wire phone jacks with conversion phone modular connector (surface-mounted).

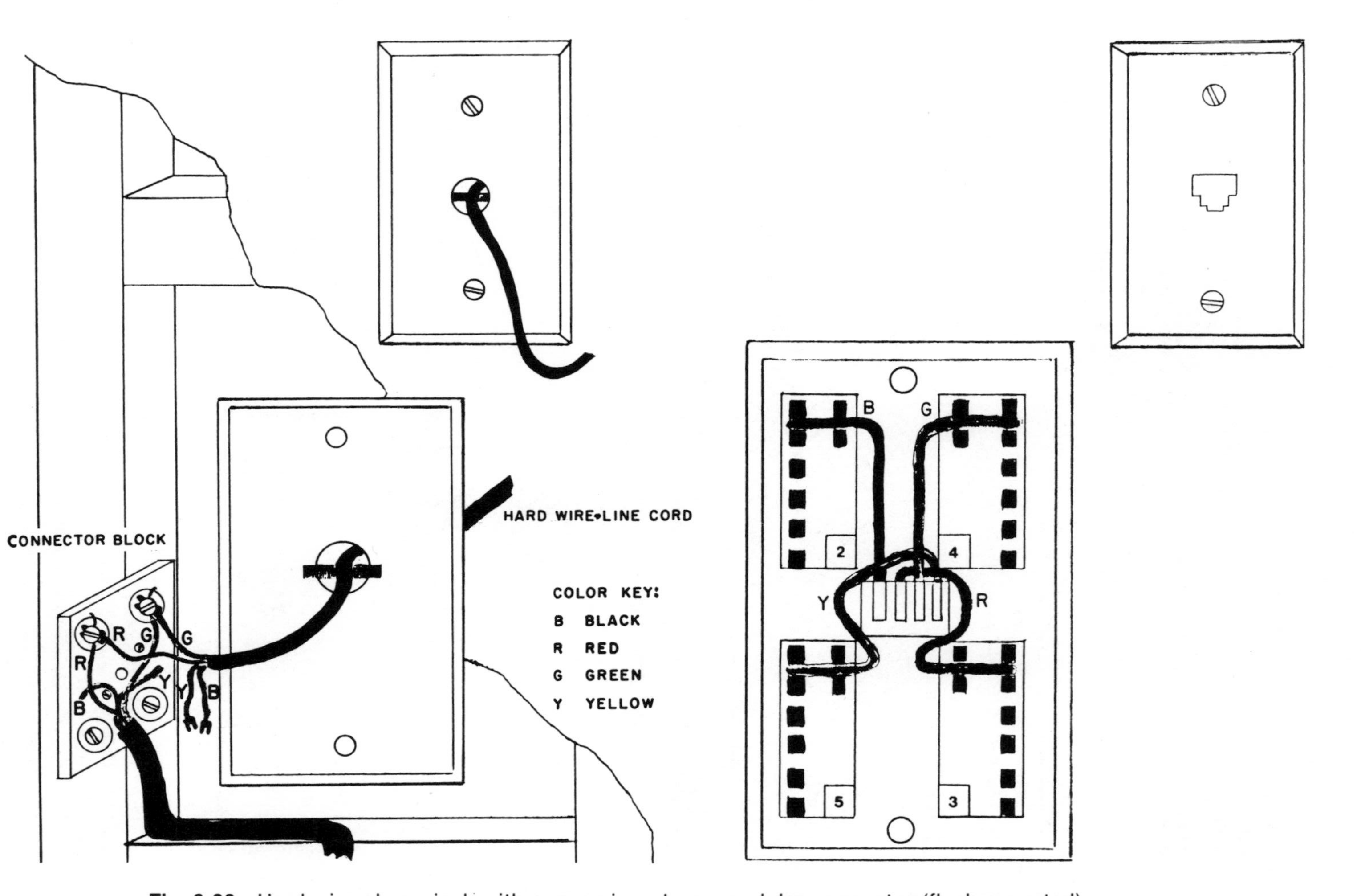

Fig. 3-38 Hard-wire phone jack with conversion phone modular connector (flush-mounted).

However, the hard-wire phone jack allows you to secure your telephone circuit.

There are two basic types of locking devices for securing telephone circuits. On rotary telephone instruments, a locking device is placed on the dialer to prevent it from rotating; on a touch tone telephone, there is a locking circuit switch that disconnects the outpulse circuit of the touch pad. When locked, these devices keep users from making outgoing calls, but incoming calls can still be answered.

These locking devices lose their effectiveness when placed on telephone instruments that are modularized because it is simple to disconnect the locked instrument and replace it with another. For touch tone instruments, it is best to have a telephone serviceperson install the lock since it requires opening the instrument and connecting it with the network and touch pad circuit, which can be difficult.

Standard Color Code for Connecting Inside Wire to Hard-Wire Phone Jack

Hard-wire phone jacks have no pin positions. Instead, they have connector terminal lugs. Thus, the standard color code connection is determined by the lead colors of the hard-wire line cord and the color code for the pair within the inside wire (Figure 3-39).

To connect the standard wiring, attach your pair, tip to one terminal lug and ring to another lug. (See Figures 3-7a, 3-7b, and 3-7c for tip and ring color code designations.) Once your pair is connected to the terminal lugs, connect the hard-wire line cord to the telephone instrument and hard-wire phone jack. Inside the telephone instrument will be a network (see Figure 1-4) and within the network will be two terminal lugs: one labeled L1 and the other L2. The L1 is your tip connection and L2 is your ring.

Converting Hard-Wire Phone Jack to Phone Modular Connector

On surface-mounted hard-wire phone jacks (see Figures 3-34 and 3-37), the 42A connector block conversion is accomplished by simply unscrewing the cover plate and attaching the leads from the conversion phone modular connector to the respective terminal lugs. This conversion connector has a mounting screw designed to screw directly into the base plate of the 42A connector block. For the connector block, the connection is the same as for the 42A connector block except the cover plate needs to be cut off with a knife since it is molded into the base plate.

The flush-mounted hard-wire phone jack (see Figure 3-38) will more than likely use a connector block to splice the inside wire to the hard-wire line cord. Simply remove the inside wire from the connector block and connect it to your conversion phone modular connector as you would for the standard connection.

Hard-Wire Wall-Mounted Telephone Converted to Phone Modular Connector

Your hard-wired wall-mounted telephone should be similar to the one in Figure 3-40, which also shows the required materials needed to convert

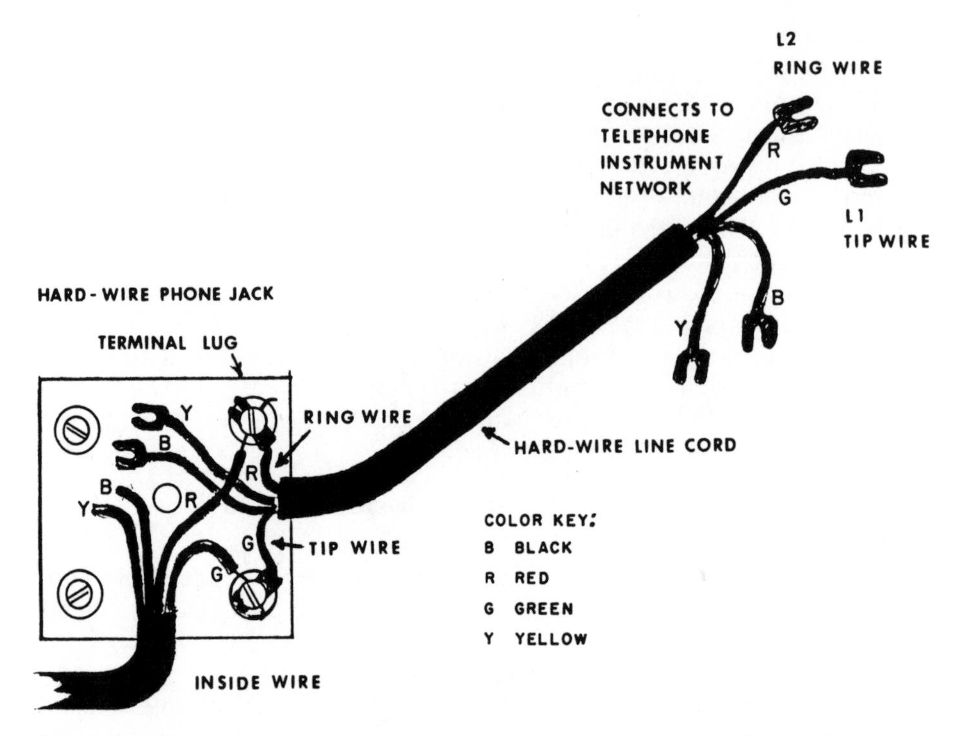

Fig. 3-39 Standard color code for connecting inside wire to hard-wire phone jacks.

it to a phone modular connector wall-mounted telephone. The inside wire is run directly into the telephone with the pair connected to the network at terminals L1 and L2. (If you can't locate the inside wire running into the instrument, check and you'll find it coming out of the wall directly behind the telephone.)

To remove the cover, push the tab at the base of the telephone instrument with a flat-head screwdriver, which will release the cover, allowing it to swing up and off. Find the pair connected at L1 and L2. After confirming which color wire goes to L1 and which to L2, remove them. Next remove the mounting screws that hold the instrument to the wall so the phone is free to set down.

To wire a flush-mounted phone modular connector (see Figure 3-32, which shows a type of flush-mounted connector that can be used as an alternative), refer back to the "Standard Color Code for Connecting Inside Wire to the Phone Modular Connector" or "Variations of the Standard Color Code for Connecting Inside Wire to the Phone Modular Connector," whichever meets your need.

Once the flush-mounted phone modular connector is in place, install the modular conversion backing plate; its leads connect directly to the network at L1 and L2 (the standard connection with the red lead, pin position 3, connected to L2 and 4-green lead, pin position 4, connected to L1). If you can't call out because the telephone is unable to break dial tone,

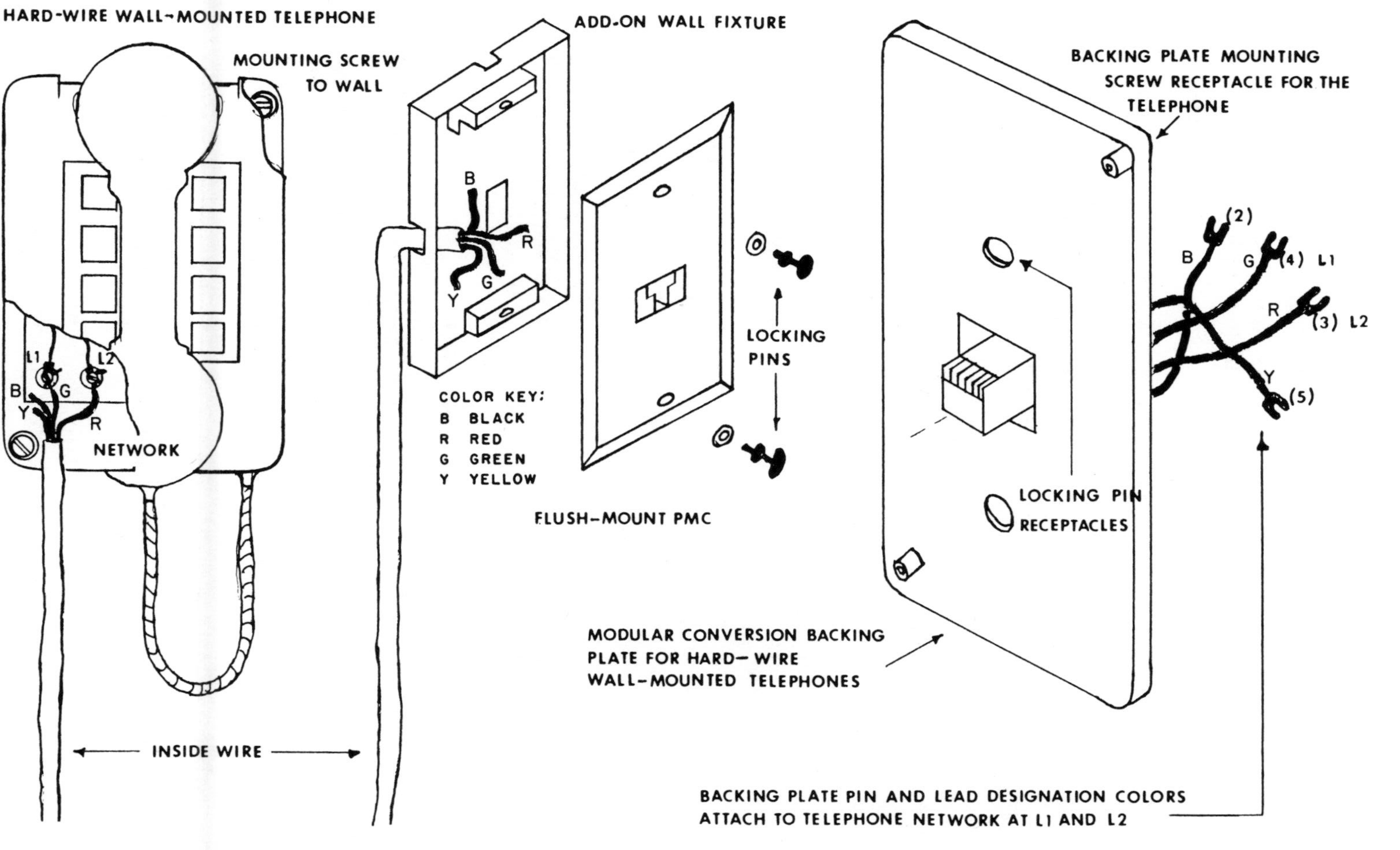

Fig. 3-40 Hard-wire wall-mounted telephones converted to phone modular connectors.

a polarity reversal is necessary: exchange the leads from the backing plate, putting the red lead on L1 and the green lead on L2. When all is working properly, install the backing plate by tightening screws from inside the instrument through to the backing plate mounting screw receptacles.

Line Cord—Modular

Line cord, the wires that connect your telephone instrument to the phone jack or phone modular connector, are a part of the national telephone standard for pin designation and lead colors. Figure 3-41 shows two modular line cords: one with six and another with four pin positions and lead colors. These modular line cords correspond to the same pin positions and lead colors as the phone modular connectors shown in Figures 3-21 and 3-22.

It is important to have the proper line cord for the job. When wiring a phone modular connector to accept a second telephone number (see Figure 3-24), the modular line cords shown in Figure 3-42 will not work because they contain three leads and two leads, which will not line up with the required pin positions (2 and 5) for wiring a second telephone number at the phone modular connector.

Modular line cords to spade leads (Figure 3-43) are necessary if you are converting a hard-wire phone jack and hard-wire line cord to a modularized phone connector. The color code and pin designations are the same as for modular line cords with the colors painted on the spades.

Modular line cords and cords to spade leads are available in 7-foot, 14-foot, and 25-foot lengths. If you need a special length, the local telephone store will usually carry the material required to make your own modular line cords.

Line Cord—Hard-Wire

Hard-wire line cords have the same lead colors as modular line cords and modular line cords to spade leads. Figure 3-44 shows two types of hard-wire line cords. One is designed for flush-mounted hard-wire phone jacks. It has a retaining bar at one end to keep it from pulling out of the small hole in the flush-mounted hard-wire phone jack. The hook on the other end catches inside so it won't pull away from the telephone instrument.

The second hard-wire line cord, for surface-mounted hard-wire phone jacks, has an eye hook that loops over a catch in the hard-wire surface-mounted phone jack to keep it from pulling out.

Installing an Extension Phone Modular Connector to a Garage Not Connected to the House

Installing an extension phone modular connector in your garage is far more preferable than using a cordless telephone because of all the inher-

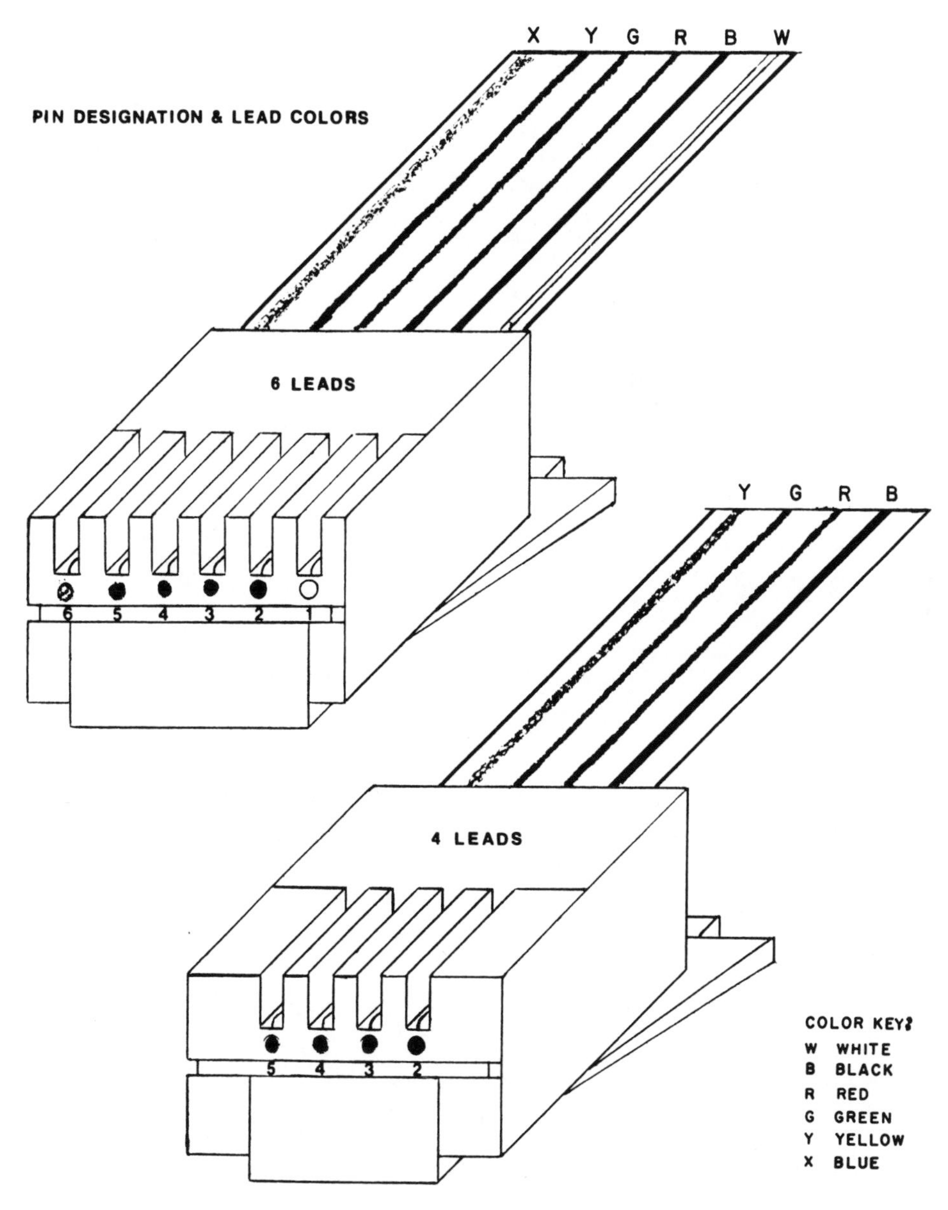

Fig. 3-41 Modular line cords (six and four leads).

ent problems associated with the cordless telephone, such as lack of privacy, false rings, and the inability to access other services.

If your telephone service is received via aerial cables, the installation of a phone modular connector to your garage can be connected as in Figure 3-45, which shows both a second aerial drop wire and house protector. The second aerial drop wire is connected to the same pair at the unprotected service terminal and then run to the second house protector, located at the garage. Since it is located before the original house protec-

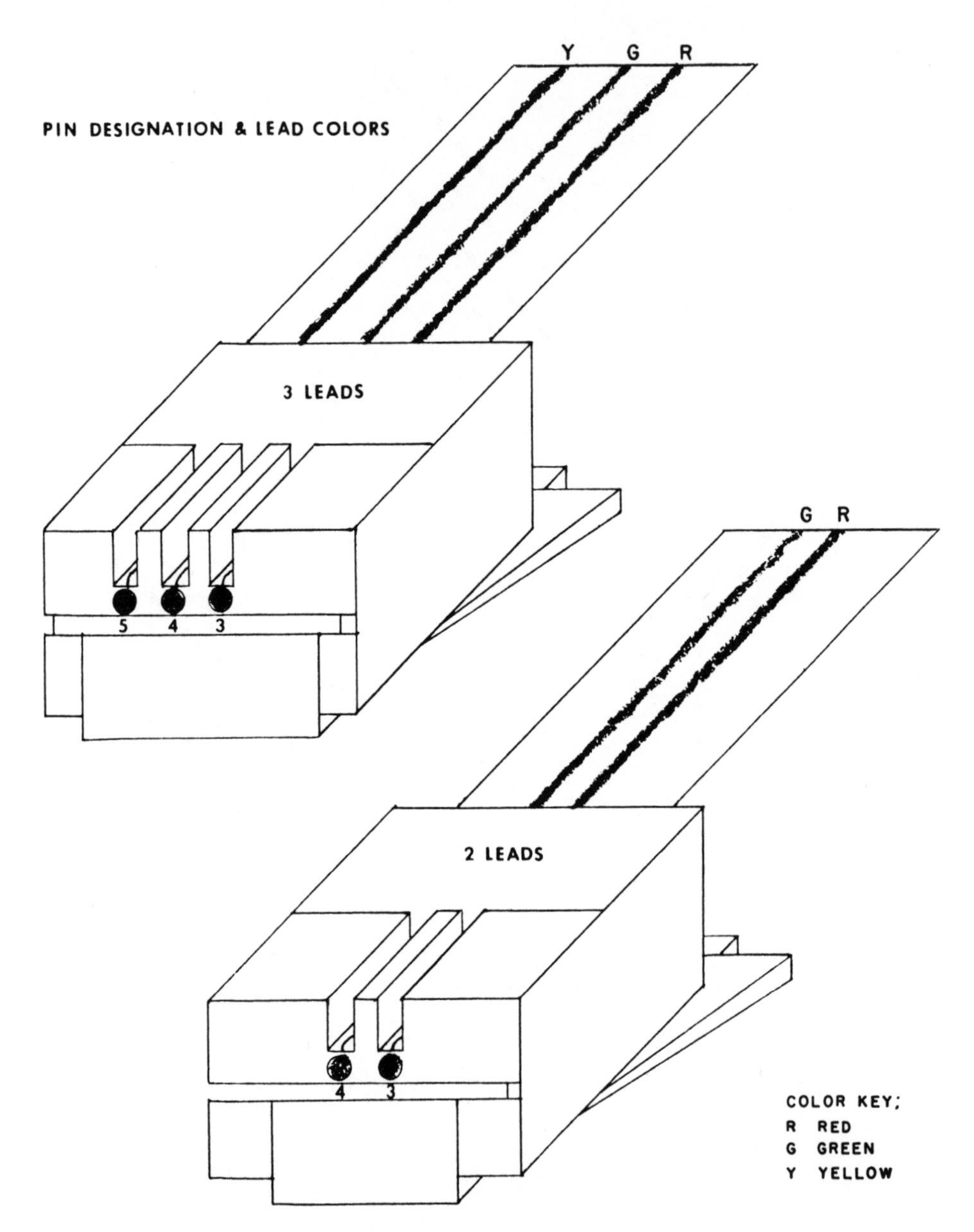

Fig. 3-42 Modular line cords (three and two leads).

tor and is not protected by its grounded circuit, the second house protector must also be grounded.

A garage that does not have an earth ground, such as a water pipe or electrical conduit, will require a ground rod. Let the telephone company know in advance that a grounding rod will be required because its personnel seldom carry them around.

By installing the phone modular connector and running the inside wire to the selected location, all that will be left to do will be for the telephone installer to place the aerial drop wire and house protector.

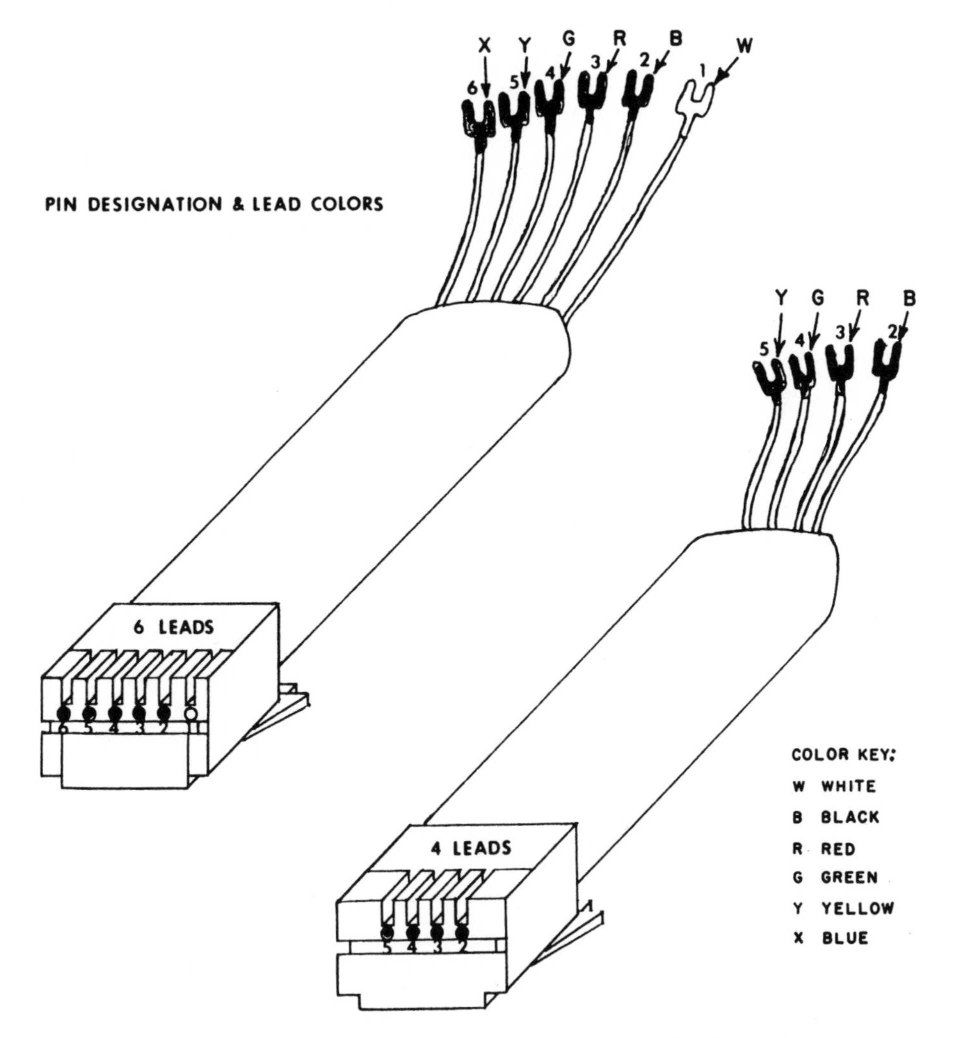

Fig. 3-43 Modular line cords to spade leads.

If your telephone service is received via buried cables, it will be necessary to place a buried drop wire from the house protector to the garage. Instead of running the buried drop wire to the patio, as shown in Figure 3-46, simply run it directly to the garage.

Installing an Extension Phone Modular Connector Outside

Installing an extension phone modular connector outside requires that it be weatherproof (Figure 3-46). If you intend to install the weatherproof connector on the house, it is all right to use inside wire to connect to it; but if you intend to install one at a patio location that is not connected to the house, you will need to place a buried drop wire.

It will be necessary to trench a path to the desired location and dig down about one foot. The buried drop wire must have three separate

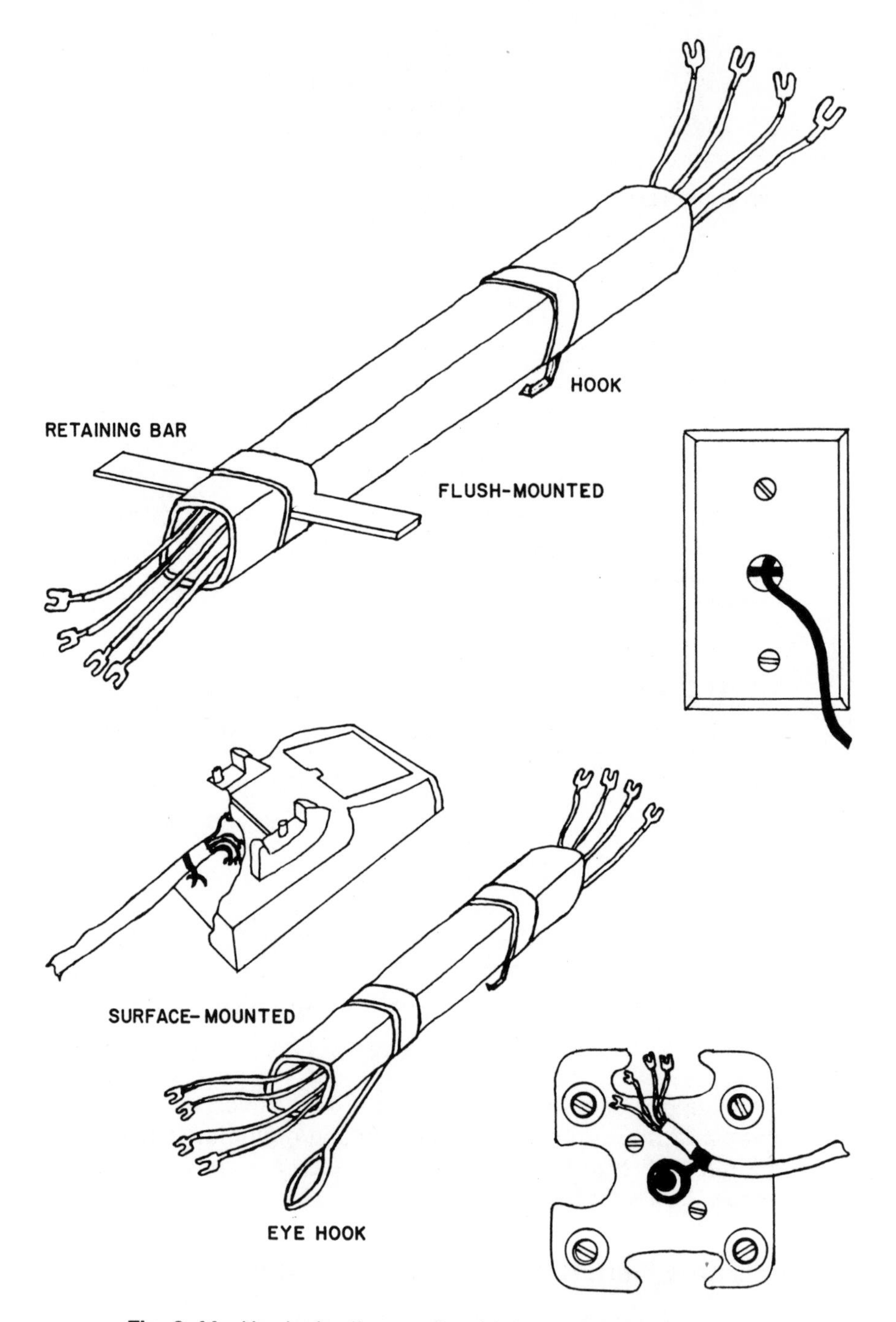

Fig. 3-44 Hard-wire line cords with hard-wire phone jacks.

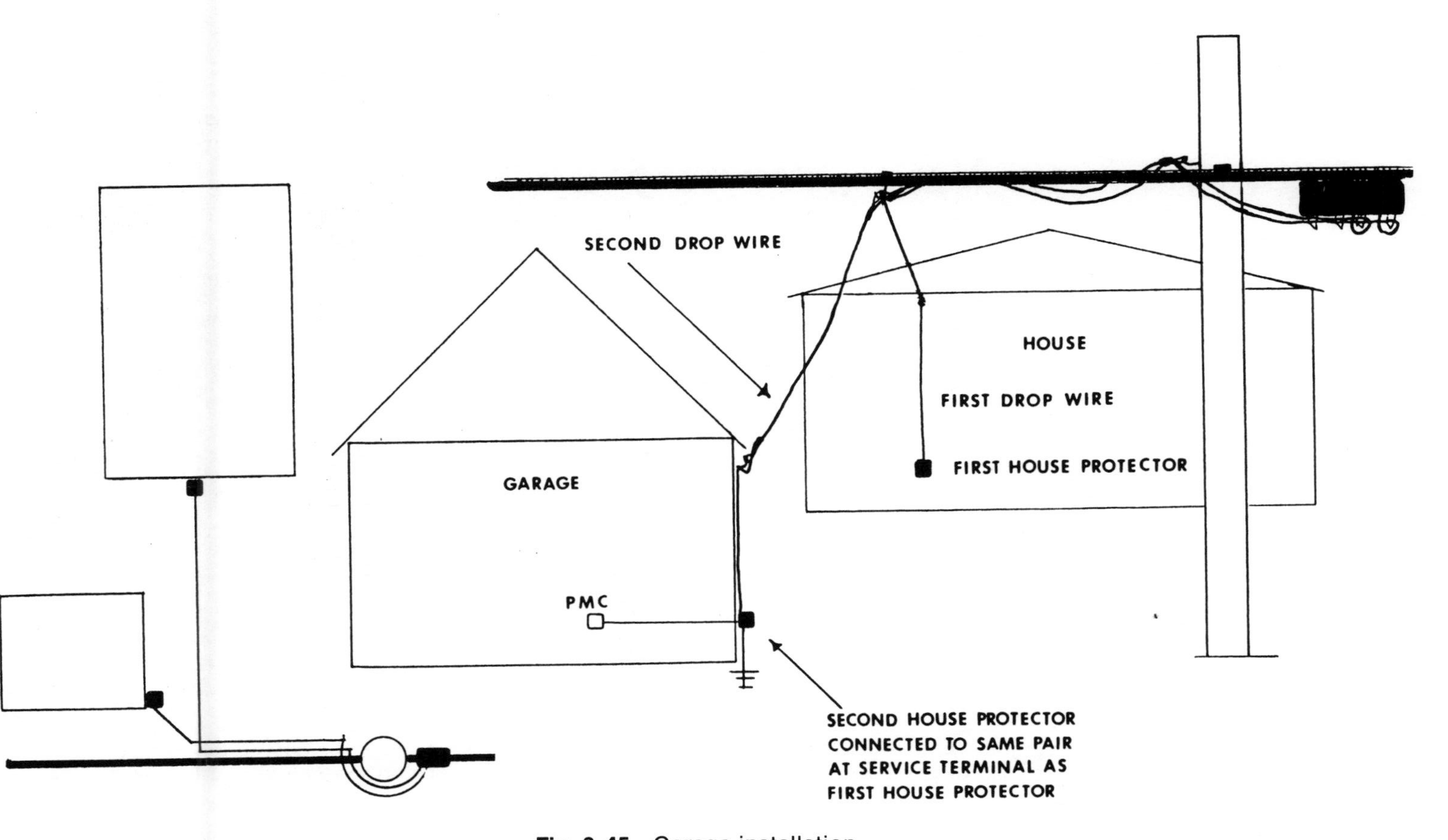

Fig. 3-45 Garage installation.

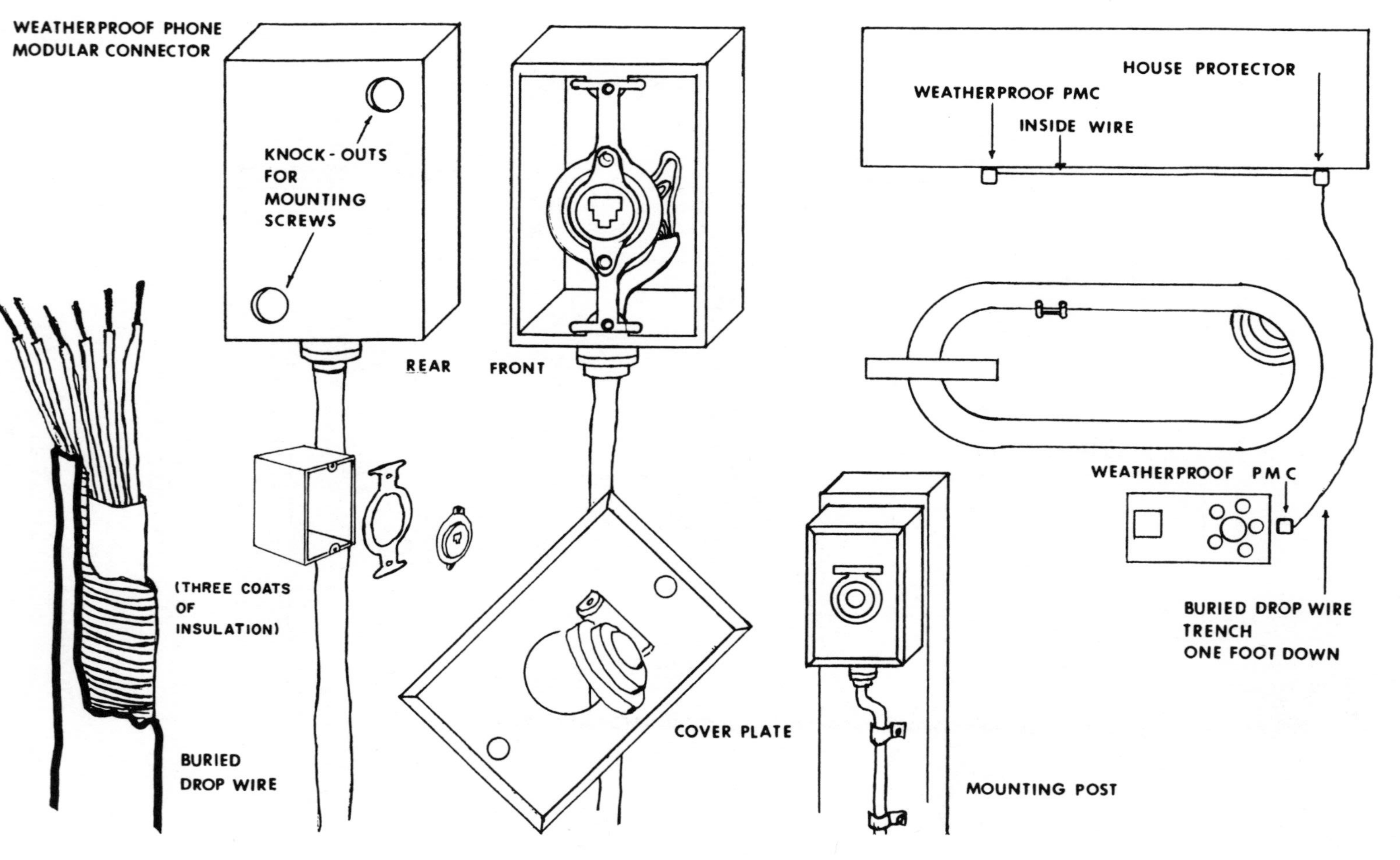

Fig. 3-46 Outside phone modular connector.

insulating coats: the outside layer a hard plastic, the next layer aluminum, and the last a soft plastic. This type of insulated cable keeps moisture from penetrating and damaging the pairs.

Materials

If you are planning to install a phone modular connector or convert a hard-wire phone jack or repair existing wiring, it may be wise to check out the Modular Conversion Kit by Gemini, which contains a good assortment of telephone materials to complete any number of telephone installation, conversion, and repair projects. Purchasing the kit usually results in major savings since you won't have to purchase items piece by piece. Also, instructions for using the kit, included on the outside of the packaging, follow standard color code procedures, provide necessary illustrations, and are easy to understand.

The Modular Conversion Kit contains the following materials:

1. One hundred feet of inside wire, two pair double insulated, marked with the standard color code (see Figure 3-7b; no mounting hardware is provided for attaching the inside wire to any type of surface.)
2. Four surface-mount phone modular connectors with four pin positions, marked with standard color code (see Figure 3-22). Two screws for each phone modular connector are included for wood surface mounting. (These phone modular connectors are recommended because they are constructed of both a base plate and a cover plate, allowing for splicing or inspection without having to remove the entire phone modular connector. Since these connectors are smaller than most, caution is required when connecting the inside wire to the terminal lugs; it is possible to short the spaded leads to its pins, which will short the pair.)
3. One surface-mount phone modular connector (conversion) designed to be used with the 42A connector block (see Figure 3-34), marked with the standard color code. One screw is included for attachment to the 42A connector block. When this connector is used in conjunction with the Five Jacks In-A-Strip multiphone modular connector, also included in the kit, it provides an excellent test and demarcation point (see Figure 4-5 in Chapter Four).
4. One surface-mount Five Jacks In-A-Strip multiphone modular connector outlet, with four pin positions. Two mounting screws included for wood surface mounting. (This connector can be used as part of your test and demarcation point when connected to the conversion phone modular connector above or as a multiphone modular connector for one phone modular connector location.)
5. Fourteen modular outlet plugs, 4 pin positions (see Figure 3-41), and one modular crimping tool. (These two materials highlight the installation concept of the kit, which is to modularize one end

of your inside wire runs and then plug them into the Five Jacks In-A-Strip, which is plugged into the conversion phone modular connector. This allows you to test each connector location quickly and accurately and to test the circuit on the central office side of the test or demarcation point. The modular outlet plugs and crimping tool can also be used to repair damaged line cords and to modularize hard-wire line cords.)

CHAPTER FOUR

Telephone Installations for Multiunit Dwellings

Although there are a few differences, the majority of telephone installation instructions and information for single-unit dwellings also apply to multiunit dwellings. The following are some variations that may be encountered when installing an instrument in a multiunit building telephone system.

Protected Service Terminal

It is at the protected service terminal that your pair is connected to the telephone company's local network (Figures I-2, I-4, and I-7 in the Introduction). Since the terminal is a shared connecting point for many other telephone circuits, the local telephone company prefers that you *not* work on your telephone circuit at its location; the chances are great that you may damage other telephone circuits and put those systems out of service. (*Note:* There is a federal regulation against tampering with telephone circuits, resulting in a high fine and/or imprisonment.)

Locating the Protected Service Terminal

Many protected service terminals are locked, prohibiting access to telephone company personnel (unless building owners or managers have allowed the company to place its own lock on the box). If you are expecting installation or repair personnel, it would be worth your time to first locate your terminal and then arrange for access.

Locating the protected service terminal usually depends on the size of the structure. For small-sized and two-story apartments or condominiums made up of 4 to 25 units, the terminal is commonly placed outside in a cabinet (Figure 4-1) or inside in a utility or washroom. For medium-sized, one-, two-, or three-story apartments or condominiums made up of 25 to 100 units, the terminal is almost always located inside in a utility room, a specifically designated telephone room, or in a basement garage (Figure

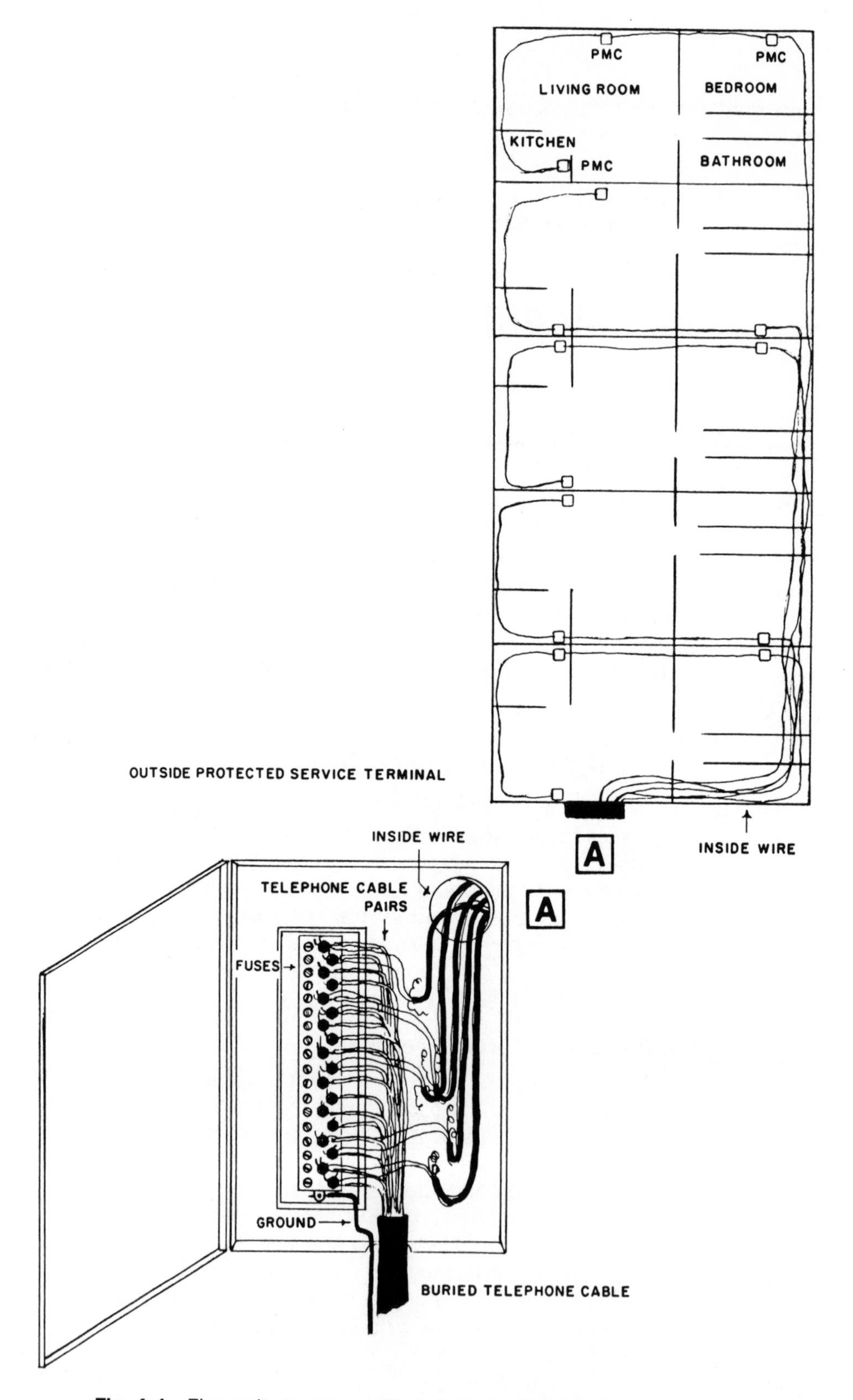

Fig. 4-1 Five-unit structure with outside protected service terminal.

4-2). In some medium-sized and in most large-sized multiunit dwellings, there will be a house cable and house terminals: the terminal, commonly placed close to your home in a hallway or utility room located on the same floor, is not fused since the protected service terminal is; and the cable is merely a way to connect cable pairs at the protected service terminal to a terminal located closer to your home. These features add flexibility by allowing for easier repair and exchange of cable pairs.

For large-sized apartments or condominiums made up of 100 units, the protected service terminal is almost always located in a telephone room specifically designed for this purpose. Large-sized units that are spread out (Figure 4-3) usually employ a management or maintenance crew that will know the location of the various locked protected service terminals. In high-rise apartments or condominiums, they are usually located next to the elevator on the first floor or in the basement. House cable and house terminals are also used frequently in high-rise structures, commonly placed next to the elevator in a utility room on the same floor where you live (see Figure 4-3).

Inside Wire

In single-unit dwellings, the inside wire is connected to the local network via the house protector, which serves as a demarcation point between the telephone company and its customer. This is also the point where inside wire can be connected for extension phone modular connectors or for testing your home telephone circuit.

Unfortunately, the multiunit dwelling has no such point.

To install your own demarcation and testing point, it is important to understand how inside wire is run through a multiunit structure (Figure 4-4), from the protected service terminal to the phone modular connectors, which are all in a series. The connector located closest to the point where the inside wire enters your home is called the number on PMC—the place to install your demarcation and test point.

Use the process of elimination to locate the number one PMC. Choose the phone modular connector you think may be the number one PMC. Disconnect the inside wire. If all the other phone modular connectors go dead, you have located the number one PMC: if not, reconnect the inside wire to the phone modular connector and try another until you find it.

Once it is located, you are ready to install your demarcation and testing point (Figure 4-5). To find the protected service terminal side of the inside wire, leave the phone modular connector hanging away from the wall. Plug in your telephone and remove one side of the wires from the phone modular connector. If you hear a dial tone, the side still connected to the phone modular connector is the proper side to the protected service terminal); if not, the side you removed is the proper one. Place a 42A connector block next to the number one PMC, as shown in Figure 4-5. Use a small piece of inside wire to connect the 42A connector block to your number one PMC. (Usually there is enough inside wire in the wall so that you can pull out a foot or so and cut off a small piece.) With this setup, you

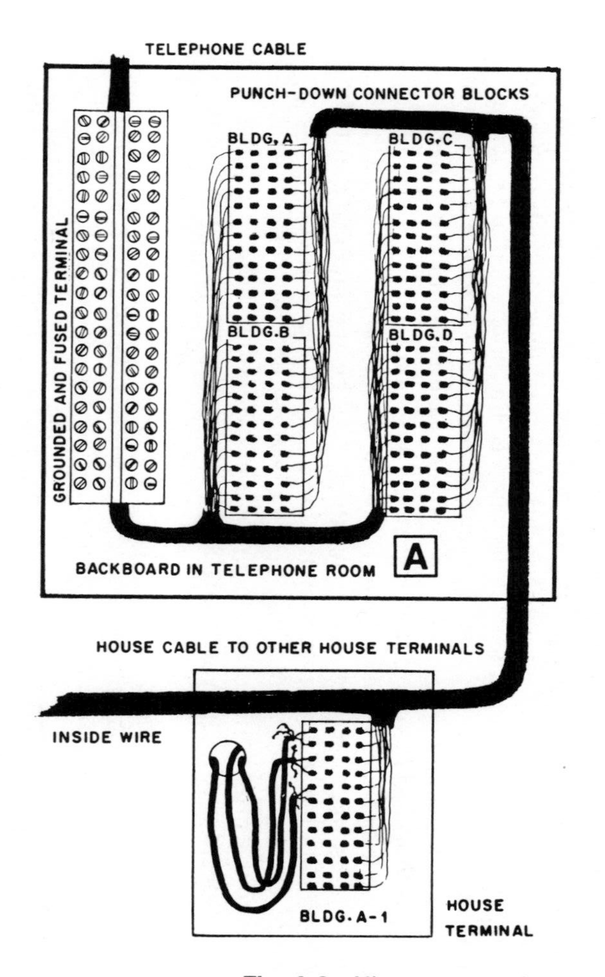

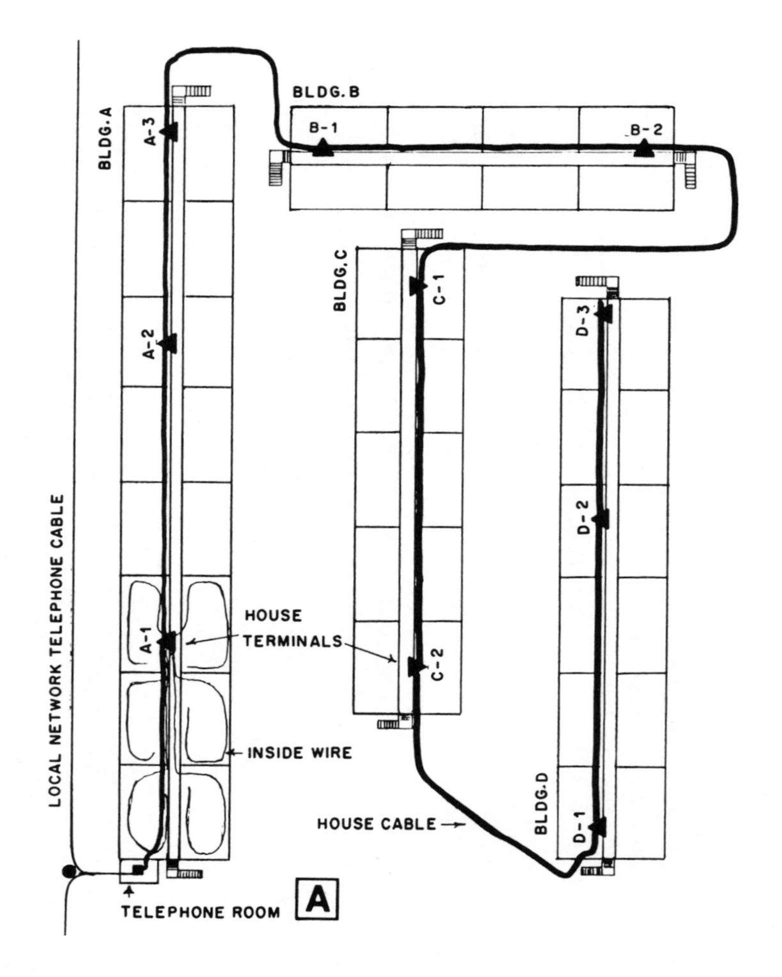

Fig. 4-2 Ninety-two–unit structure with inside protected service terminal.

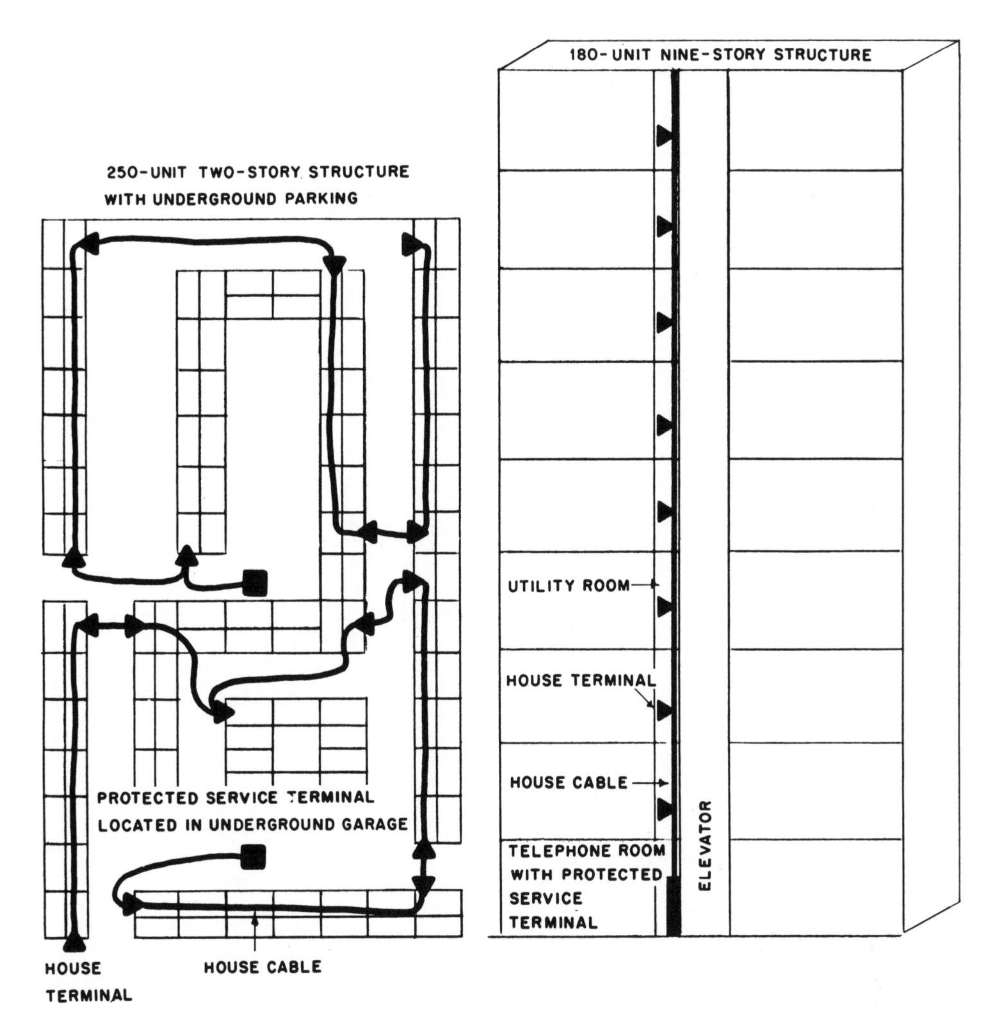

Fig. 4-3 Large multiunit dwellings with inside protected service terminals.

now have an easy place to test your home telephone system along with a point where you can add inside wire for additional or extension phone modular connectors.

Figure 4-6 shows a typical multiunit dwelling inside wire run when the inside wire is of the one pair type, which is the key in determining if your home has this type of split arrangement. Inside the protected service terminal will be two inside wire pairs connected to the same terminal lugs, a good arrangement for adding an additional telephone number because inside wire A or B can be used for the second number, eliminating the need to install extra inside wire or a phone modular connector. For this arrangement to work, however, you must be satisfied with the existing connector location, and you must be willing to do without your current first telephone number on inside wire run A or B; otherwise, it will be

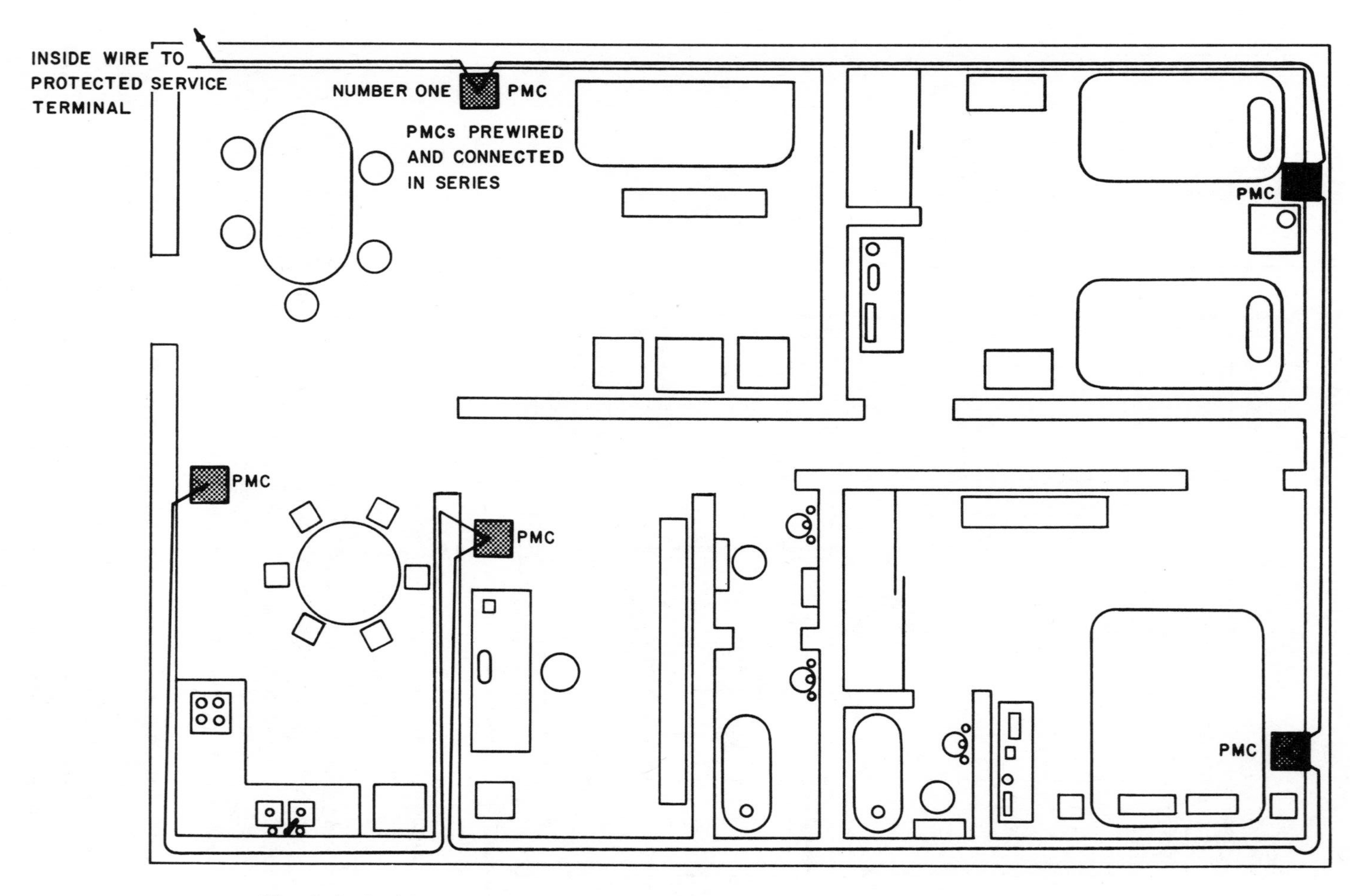

Fig. 4-4 Inside wire run and PMC connection for typical multiunit structure.

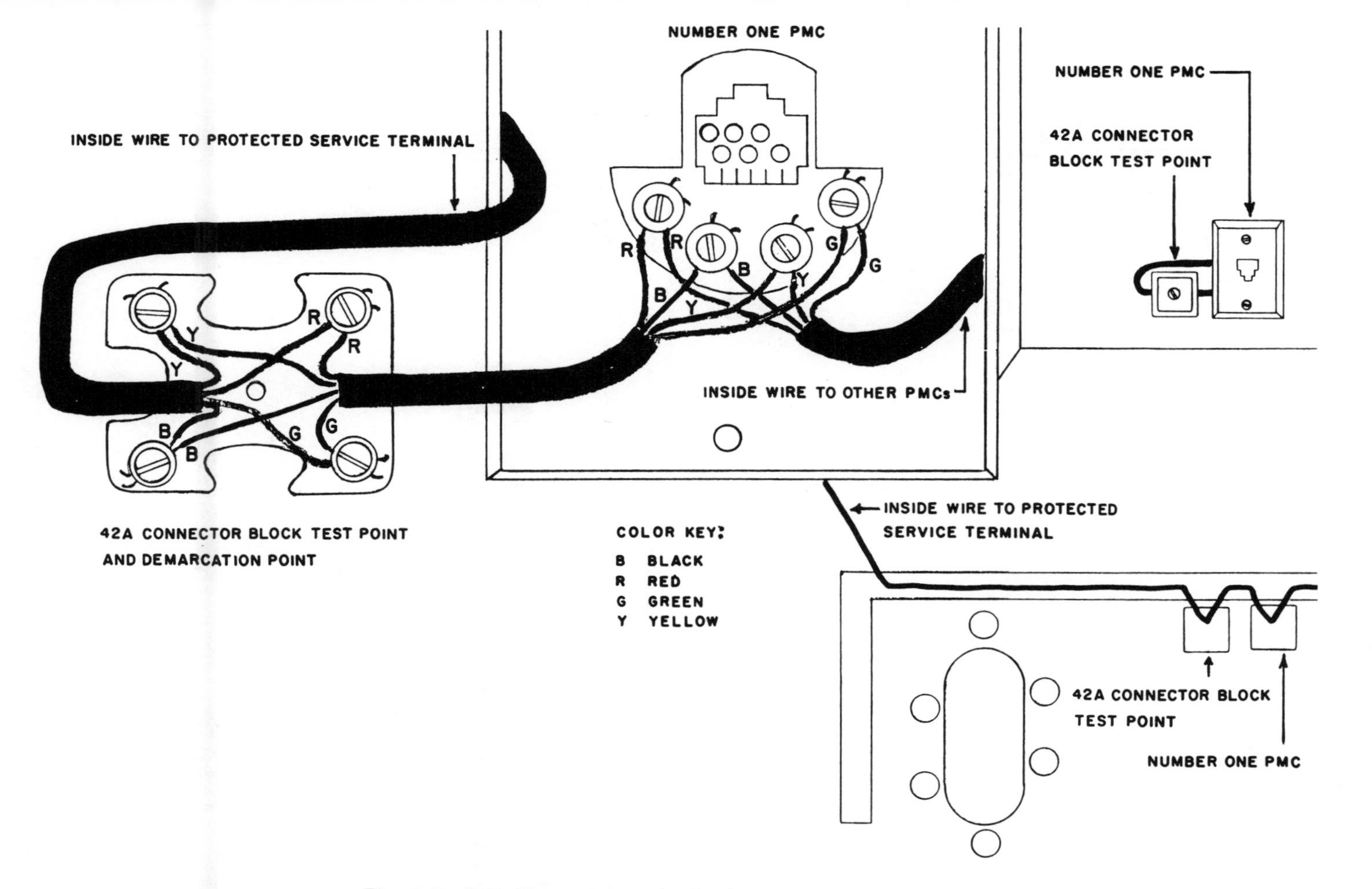

Fig. 4-5 Installing a test point for in-series PMC hookups.

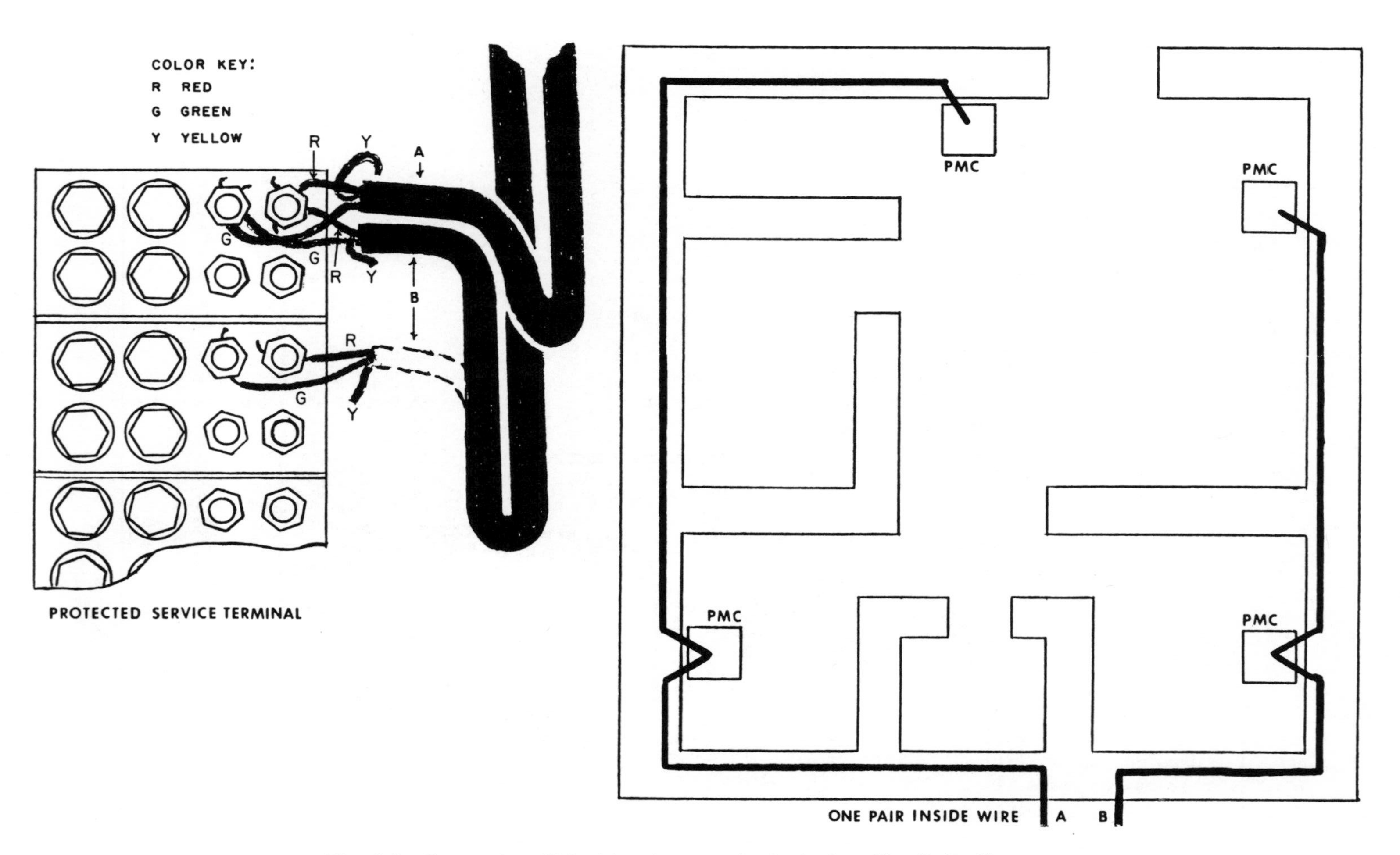

Fig. 4-6 One-pair split inside wire runs for typical multiunit dwelling.

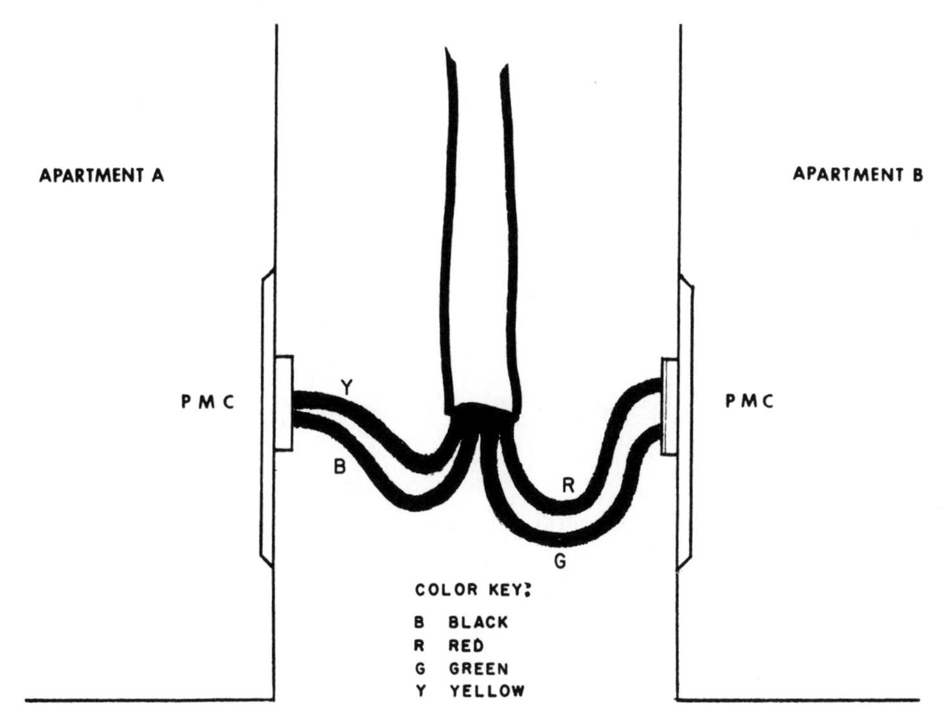

Fig. 4-7 Two apartments sharing the same inside wire.

necessary to install your own extension phone modular connectors from existing connectors on the inside wire run A or B, whichever you choose (see Figure 3-16 in Chapter Three).

When two separate apartments use the same inside wire, there is no extra pair to replace one in apartment A that goes bad (Figure 4-7). This type of arrangement would also be a problem if you intend to have more telephone numbers than available pairs. Also, to replace damaged inside wire or additional wire from the protected service terminal to your apartment, the local telephone company will require a Building Owner's Consent Form since the building you live in is owned by a third party. Owners and associations almost always have rules against placing wires along their buildings; you will save yourself a great deal of trouble by having the form signed ahead of time. Figure 4-8 shows a typical consent form.

Prepared by	*Date*
Service Order Number	
Telephone Number	

Building Owner's Consent

As owner of the building(s) located at

____

(*Street Address*) (*City*)

I hereby consent to the installation by

____

(*Name of local telephone company*)

consisting of wires, cables, and other necessary fixtures and attachments in or on said building(s), other than in the conduits and other facilities specifically provided for such purpose, in that portion of said building described as follows: ______________________________

____

____

____

____

______________________________ ______________

(*Owner's Signature*) (*Date*)

______________________________ ______________

(*Street Address*) (*City and State*)

Fig. 4-8 Building owner's consent form. *Note:* Make three copies: one for the telephone company, one for the building owner, and one for your records (all signed).

CHAPTER FIVE

Telephone Repair

With the information in this chapter, you will be able to troubleshoot and repair your own home telephone system. Also included in this chapter are instructions for special repair situations along with tips on how to ensure proper and efficient service from your local telephone company.

The Telephone Circuit in Trouble

There was a time when you could simply call 611 to have a repair technician dispatched to your home—at no charge to you—to fix whatever was wrong with your home telephone system. However, since the deregulation of the telephone industry, the local telephone company has relinquished control over certain areas within your home telephone system; customers are now billed for service in those areas.

Before beginning any telephone repair work, you must isolate the problem by examining the telephone circuit as it travels from your instrument to the local telephone company's central office.

The telephone circuit from your instrument to the central office covers a variety of areas of responsibility (Figure 5-1). The telephone instrument is your responsibility, whether you rent it or buy it. If the trouble is isolated to the instrument, the repair technician will not fix it—but you will be charged for the visit. (Some companys do repair telephones at no charge for the elderly or disabled.) The technician may repair phone modular connectors and inside wires at an extra charge. (Some companies now charge by the hour for this service.)

The area of demarcation is that point in your home telephone system where the telephone company takes full responsibility, with no additional repair charges to you. In single-unit dwellings, this point may vary depending upon your local telephone company's policy. Some companies take full responsibility from the house protector all the way back to the central office; others take responsibility at the unprotected service terminal, leav-

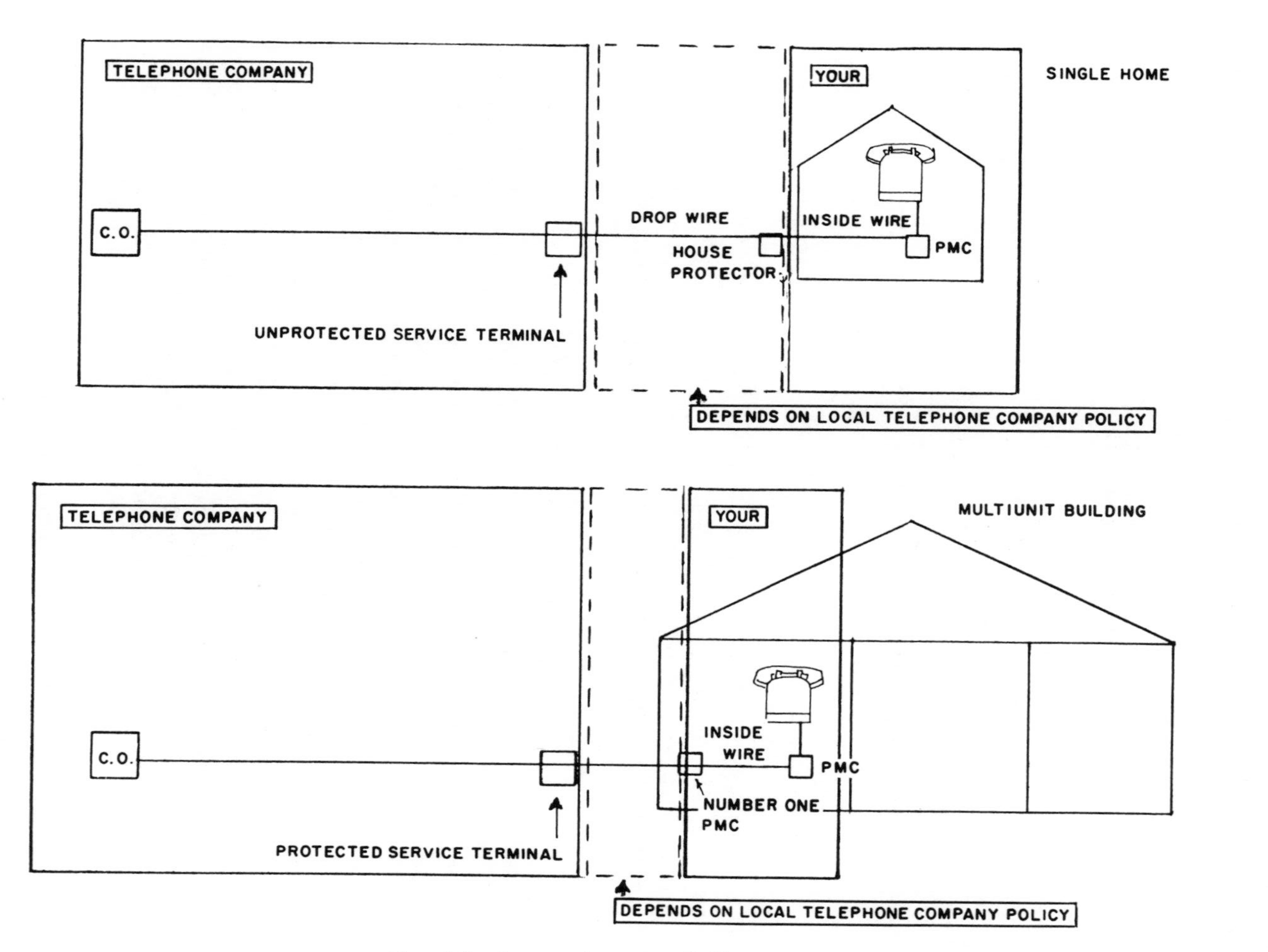

Fig. 5-1 Areas of responsibility.

ing the house protector and drop wire to you, which means you will be struck with additional repair charges because the telephone company's repair technicians are the only personnel authorized to work in this area. With buried drop wire, you can have someone else replace it, but they cannot connect it to a buried service terminal.

With multiunit dwellings, some telephone companies use the protected service terminal as the demarcation point and make additional charges for any trouble that is isolated to either the house cable or inside wire that goes beyond it; others recognize that you have no means of isolating trouble to house cable and inside wire and will repair trouble up to the number one PMC (see Figure 4-4 in Chapter Four) at no additional charge. Since the area from the demarcation point back to the central office is the responsibility of the local company and troubles isolated to that area are repaired for free, it may be wise to install your own demarcation and test point (see Figure 4-5 in Chapter Four).

By isolating trouble past the demarcation point, you will not need to wait at home for the technician since all the repair work will be performed in the field.

Automatic Testing for Telephone Circuit Trouble

In an attempt to provide customers with better repair service and to continuously monitor telephone circuits for trouble, telephone companies that have state-of-the-art testing equipment have in operation a testing computer that automatically monitors and tests all the central office's telephone circuits.

When coming across a circuit that is indicating trouble, the computer flags that circuit and notifies switching crews in the central office, who disconnect it from the switching equipment (this is known as "killing the pair"). The computer then notifies the dispatching department, where a trouble ticket is generated and routed to a repair technician, who isolates the trouble and repairs it.

The automatic testing is usually performed at low-traffic periods during the late evening hours. If you get a mysterious ring from your telephone every evening at the same time yet no one answers, it may be that the automatic testing computer is checking the circuit and your pair has a polarity reversal.

With automatic testing, it is possible for your local telephone company to know about a problem with your circuit that you may not even be aware of. For example, you may be able to place outgoing calls yet, at the same time, not receive calls. This problem may be caused by a corroded phone modular connector, which will short your pair when the electrical voltage from an incoming call jumps across the pins within the phone modular connector and disconnects the incoming call the second you answer the telephone; thus, you are likely to think the caller dialed a wrong number and then simply hung up when you answered.

In local service areas that have automatic testing equipment, it is best not to leave your telephone instrument off the hook for extended periods of time because the computer will see this as trouble and disconnect your telephone circuit from the central office. By leaving the telephone off the hook, you are also tying up switching equipment. To avoid nuisance calls, or for those times when you just don't want to be disturbed by a ringing phone, simply unplug your telephone instruments.

Five Basic Types of Telephone Circuit Trouble

By examining the five basic types of telephone circuit trouble, you will have some idea as to their effect on your telephone circuit, which, in turn, will help you to troubleshoot and isolate problems needing repair.

The five basic troubles are opens, grounds, shorts, crosses, and equipment failures, and it is possible to have a combination of troubles at any one time.

Open Telephone Circuit

With the open telephone circuit, the entire circuit is not affected. It is possible to have three types of opens (Figure 5-2).

With both the ring and tip wires open, no dial tone or line voltage will be detected on the telephone equipment side of the open circuit. With only the ring wire open, no dial tone will be detected unless you connect one lead of the test set (see Figure 5-6) to the tip wire and connect the other lead to the ground. You will have a dial tone and line voltage and be able to call out; however, there will be a loud humming noise on the circuit. With only the tip wire open, no dial tone or line voltage will be detected on the telephone equipment side of the open circuit.

The voltmeter (see Figure 5-7) can be used to test for an open telephone circuit since the lack of line voltage will be indicated when its test leads are connected to the tip and ring wires. The tone set and probe (see Figure 5-9) can also be used: when the open is between the two points being tested, no tone will be detected at the end where you are trying to locate the tone.

In service areas with automatic testing, the open telephone circuit will not be disconnected from the central office switching equipment since the entire telephone circuit is not affected. However, the lack of a dial tone and line voltage does not necessarily mean you have an open circuit. In order to isolate the trouble to an open circuit, it will first be necessary to test for shorts and grounds to ensure that neither is affecting the circuit.

Grounded Telephone Circuit

The grounded telephone circuit, which affects the entire circuit (Figure 5-3), falls into three categories.

When only the tip wire is grounded, dial tone will be detected along with proper line voltage, unless the telephone circuit has been disconnected because of automatic testing. Although there will be a humming sound with loud static, it will still be possible to call out. A tip ground can be detected with the ohmmeter (see Figure 5-7) by placing one test lead on

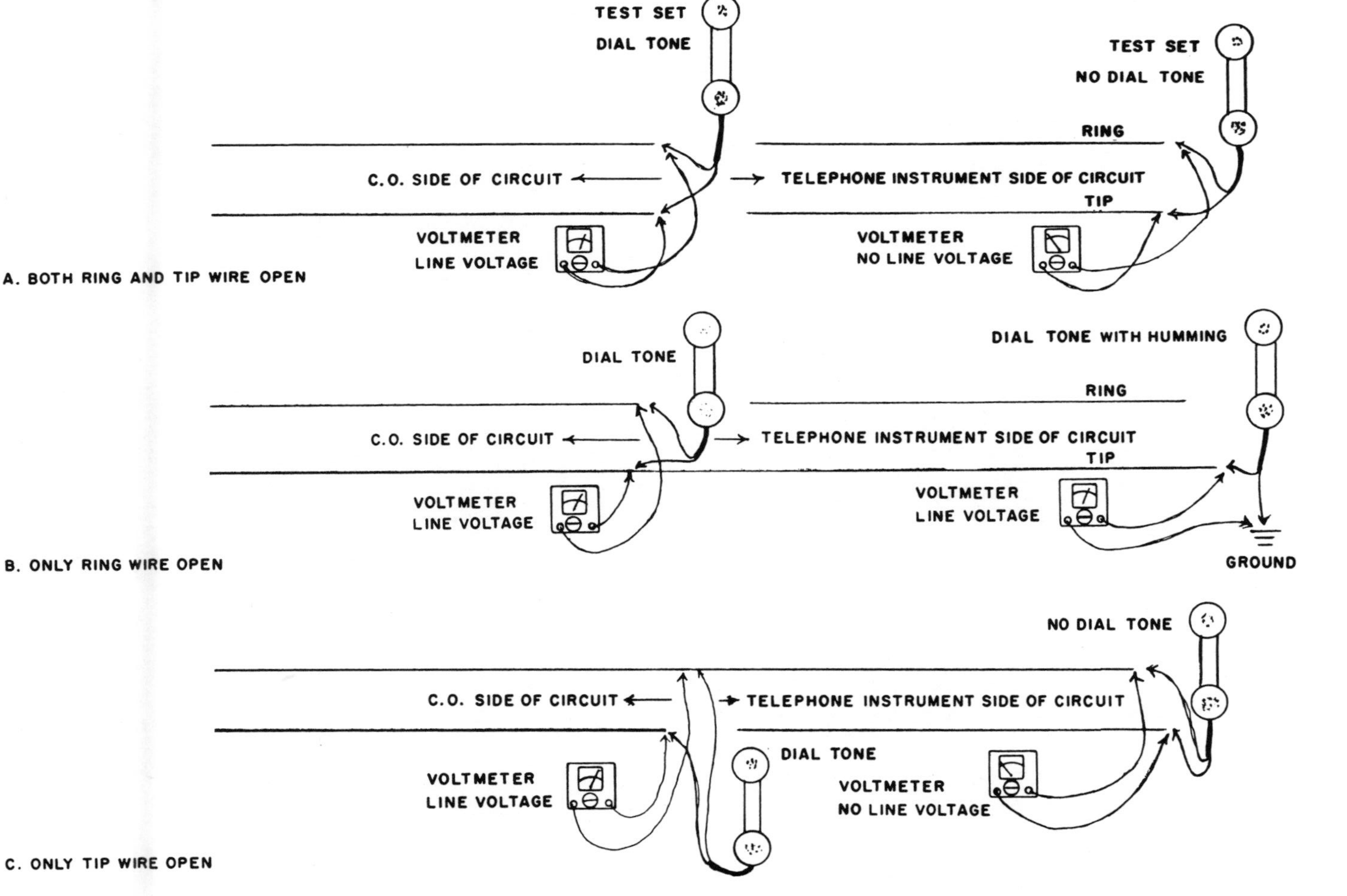

Fig. 5-2 The open telephone circuit.

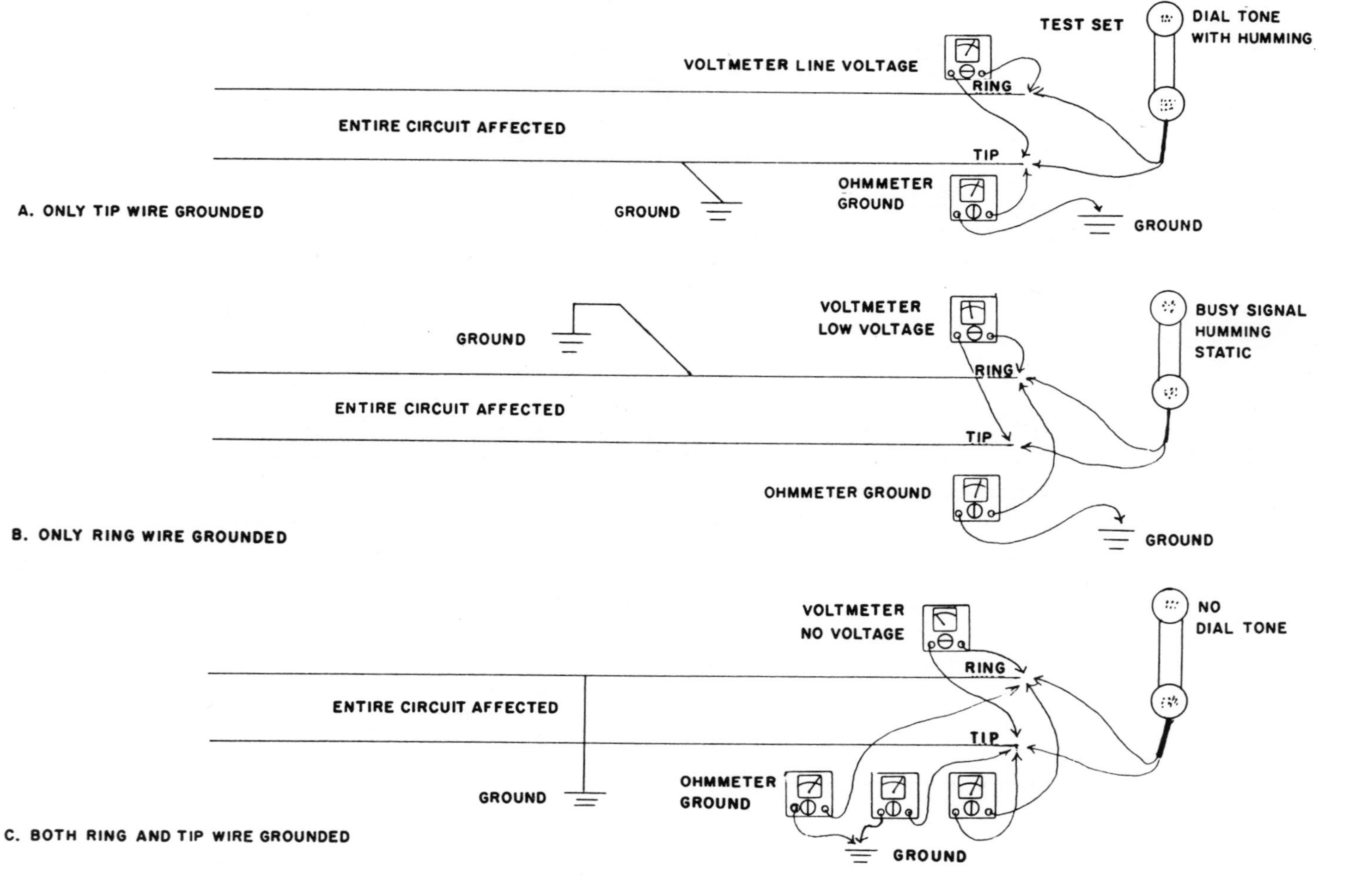

Fig. 5-3 The grounded telephone circuit.

the tip wire and the other test lead on the ground; the needle should swing all the way to the right.

When only the ring wire is grounded, a continuous busy signal will result, and line voltage will be lower than it should be. The telephone circuit will have a humming sound with loud static. A ring ground can be detected with the ohmmeter by placing one test lead on the ring wire and the other test lead on the ground; the needle should swing all the way to the right.

With both the ring and tip wire grounded, no dial tone or line voltage will be detected. The telephone circuit will also have a short. This type of telephone trouble can be detected with a volt and ohmmeter: the voltmeter will detect no line voltage, and the ohmmeter will independently detect tip and ring grounds.

Shorted Telephone Circuit

The shorted telephone circuit affects the entire circuit (Figure 5-4), and occurs when the ring and tip wires make contact with each other. When the telephone circuit is shorted, no dial tone or line voltage will be detected. If automatic testing has disconnected the circuit because of a short, use the volt and ohms meter to detect it by placing one test lead on the tip wire and one on the ring wire. Set the selector switch to 60 volts DC; no voltage should be indicated. Next set the selector switch to ohms X100; the needle should swing all the way to the right.

Crossed Telephone Circuit

The crossed telephone circuit affects every circuit it is crossed with (Figure 5-5) and occurs when either the tip or ring wire in the pair makes contact with either the tip or ring wire in the pair of another telephone circuit.

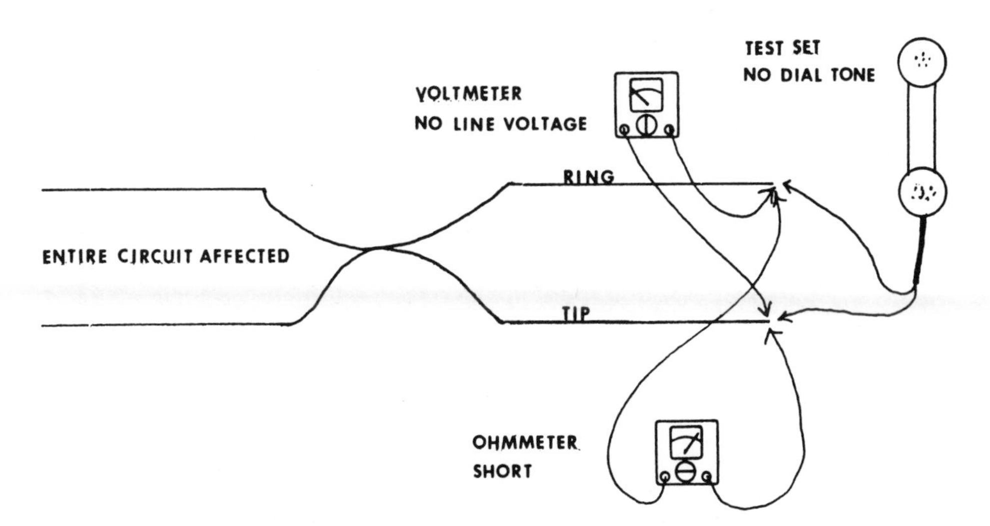

Fig. 5-4 The shorted telephone circuit.

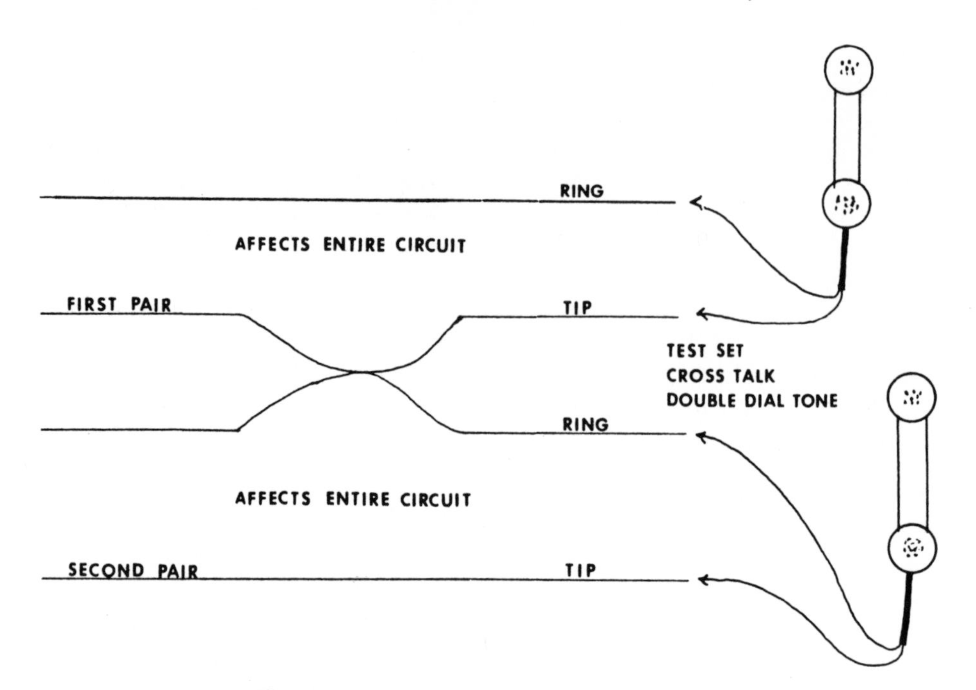

Fig. 5-5 The crossed telephone circuit.

The exact effects of a crossed circuit depend on the type of cross that exists: if two tip wires touch, if tip and ring wires touch, or if two ring wires touch. The general effects include hearing another conversation, hearing two distinct dial tones, hearing a humming noise, or hearing a combination of the above.

Crossed telephone circuit trouble almost always is isolated past the demarcation point toward the central office. However, if you have more than one telephone number, it is possible that these circuits could be crossed. To check, simply pick up the second instrument. You will know the circuit is crossed if you hear any of the telltale warning noises.

Equipment Failures

The effects of equipment failures, which may or may not affect the entire telephone circuit, can be anything from misdialing to no dial tone. Equipment you may have includes telephones, answering machines, modems, auto dialers, memory dialers, and alarm control units. If the problem clears once a piece of equipment is disconnected, then you have isolated the trouble to that particular piece of equipment; if not, check the station wiring for trouble and continue the search, which could lead straight back to the central office.

Test Equipment

Test equipment becomes an important factor in isolating telephone circuit trouble when you need to do it in the least amount of time. The following examines the equipment most commonly used by telephone company repair personnel and provides instructions on how to test for shorts, grounds, and opens. Tips on constructing your own test equipment out of materials already at hand—and at minimum expense—will also be detailed.

Test Set

The test set is a small telephone that looks like a handset; a telephone company serviceperson carries it on his or her utility belt or uses it at the service terminal (Figure 5-6).

The test set operates like any other basic telephone instrument with but three exceptions. First, it does not have a modularized line cord; instead, it has insulated alligator clips that can be connected to a telephone circuit at terminal lugs and that are also able to penetrate inner cable pair insulation, providing a connection to a telephone circuit without stripping wire. Second, the test set has a monitor function that, when switched on, allows the user to listen to the telephone circuit without receiving a dial tone (similar to the mute function). The monitor function also allows the user to connect to a telephone circuit that is in use without interrupting the conversation. Third, the test set has no ringer built into it so the user must listen at the receiver to hear a clicking sound, which indicates an incoming call.

For the home user, the test set will help isolate telephone circuit trouble when the circuit is opened up to detect dial tone on the central office side of the circuit and tone on the telephone instrument side.

To make your own test set, you need an inexpensive telephone instrument (they are lightweight and carry everything in the handset), a phone modular connector (preferably a surface-mounted type with terminal lugs), two or three feet of inside wire, and two insulated alligator clips (see Figure 5-6).

Connect the alligator clips to the inside wire then connect the inside wire to the standard 3-red and 4-green pin positions at the phone modular connector.

Volt and Ohms Meter

The volt and ohms meter, which can be used to determine either volts or ohms by the flick of a switch (Figure 5-7), indicates whether or not the telephone circuit is carrying the proper circuit voltage. To test (see "The Test Point," p. 153, to find out where to test the circuit), set the selector switch on the volt and ohms meter to 60 DCV (60 Direct Current Voltage). Take the positive (+) test lead and place it on the tip wire of your telephone pair and then take the negative (−) test lead and place it on the ring wire (refer back to Figure 3-7a through 3-7c to find out which is ring and which is tip). The needle in the meter window should indicate 48 Volts DC, though if someone is talking at the time, it will register only 24 Volts DC.

In order to test for shorts, the telephone circuit must first be disconnected so that no voltage is on the circuit. Set the selector switch to ohms X 100. Take one test lead and place it on the ring wire of your telephone

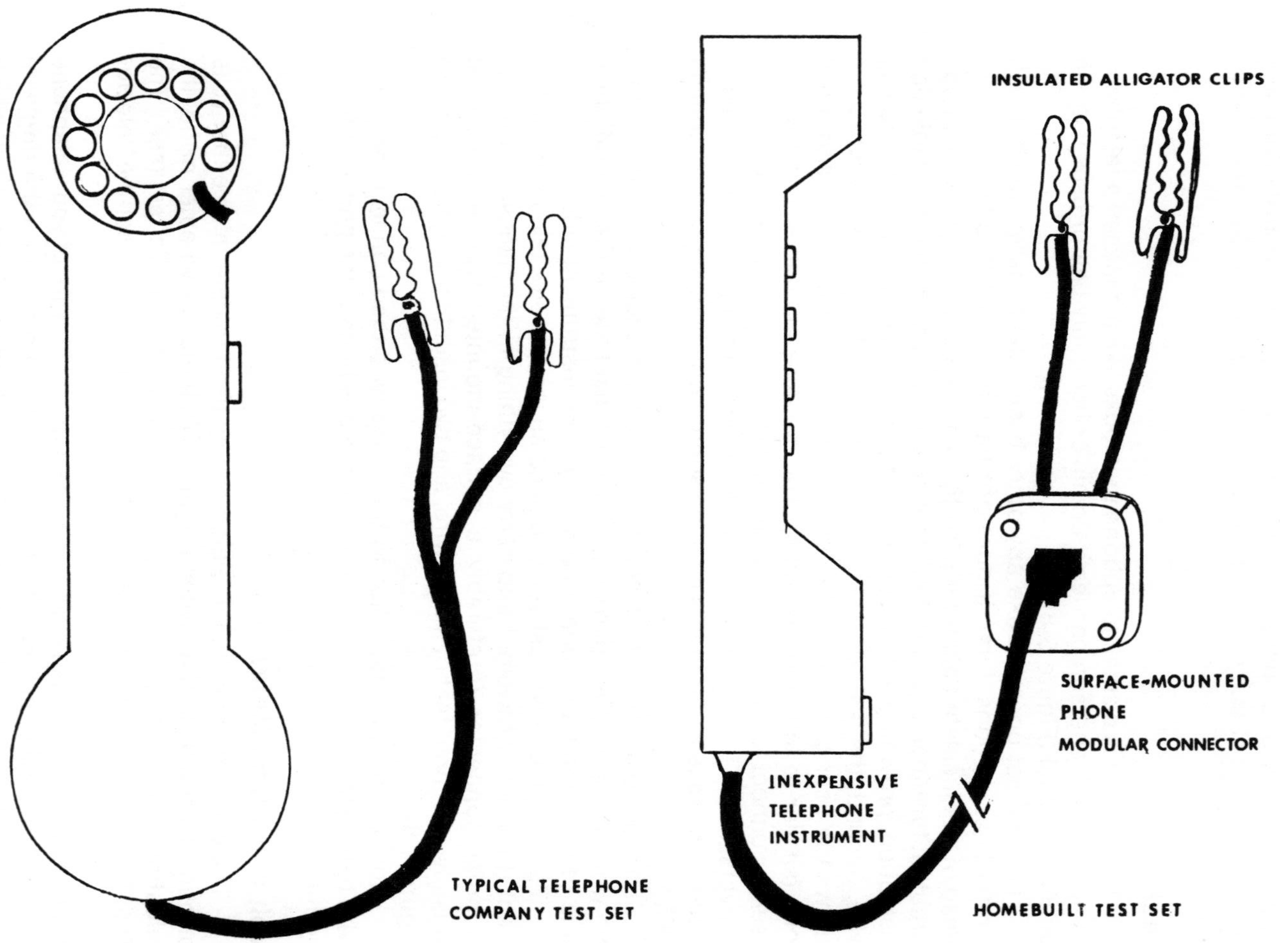

Fig. 5-6 The test set.

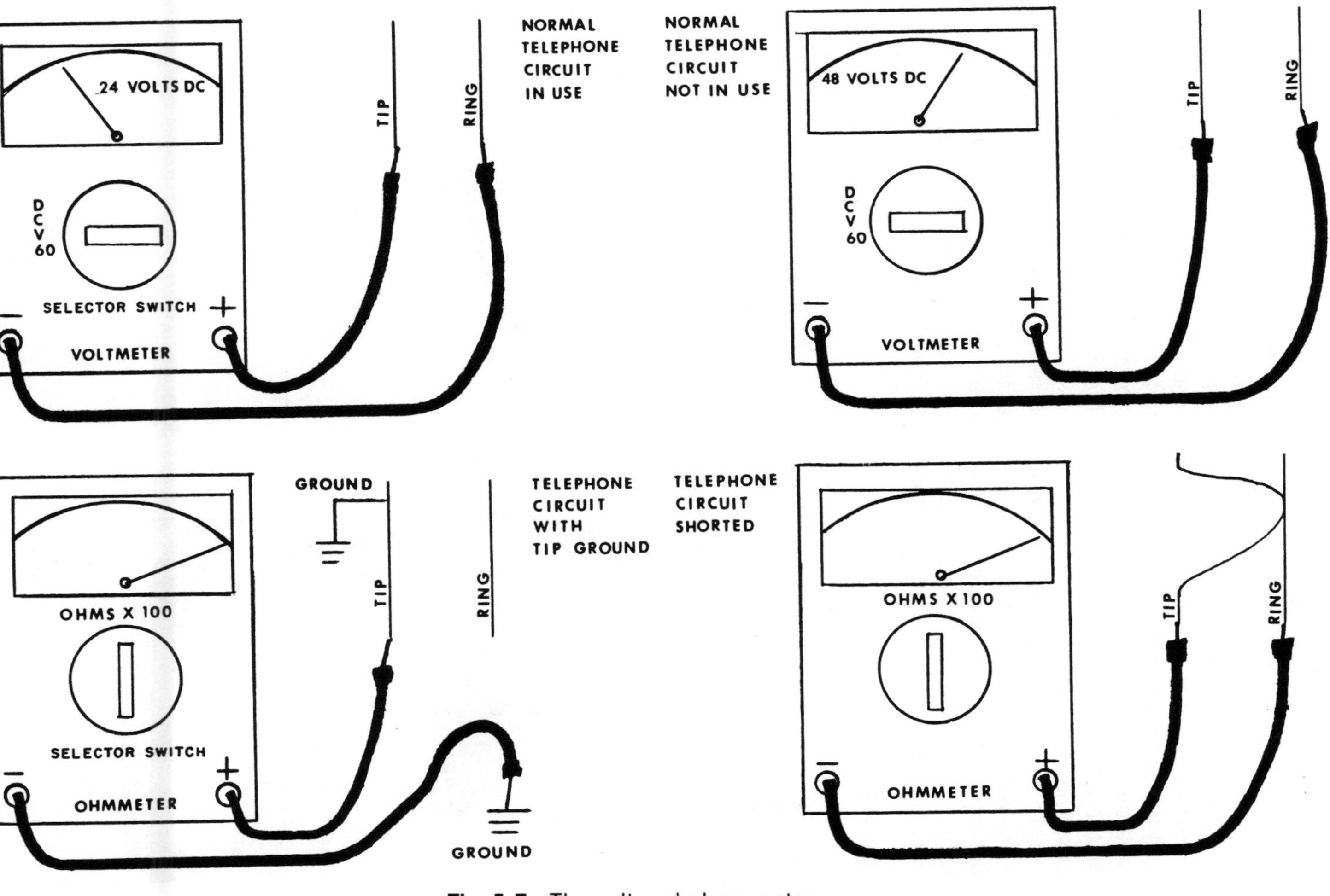

Fig. 5-7 The volt and ohms meter.

pair and then take the other test lead and place it on the tip wire. If there is a short, the needle will swing all the way to the right and stay there until you release one or both test leads (see Figure 5-7).

The volt and ohms meter also indicates whether or not the telephone circuit is grounded. Instead of connecting the ohmmeter leads to the tip and ring wires of the pair, connect one lead to the grounded terminal lug at the house protector and the other lead to the tip wire. If the needle swings all the way to the right, a grounded tip wire is indicated. Repeat the process with the ring wire.

To make your own short and ground tester, you will need two wire leads with insulated alligator clips on the ends, a buzzer, a 9-volt transistor battery, and a small amount of tape (Figure 5-8). To use your assembled tester, follow the testing instructions for a short or ground; instead of a needle swinging to indicate trouble, the buzzer will sound.

Tone Set and Probe

The ideal test equipment for identifying telephone cable pairs is the tone set and probe (Figure 5-9). When used in conjunction with a test set,

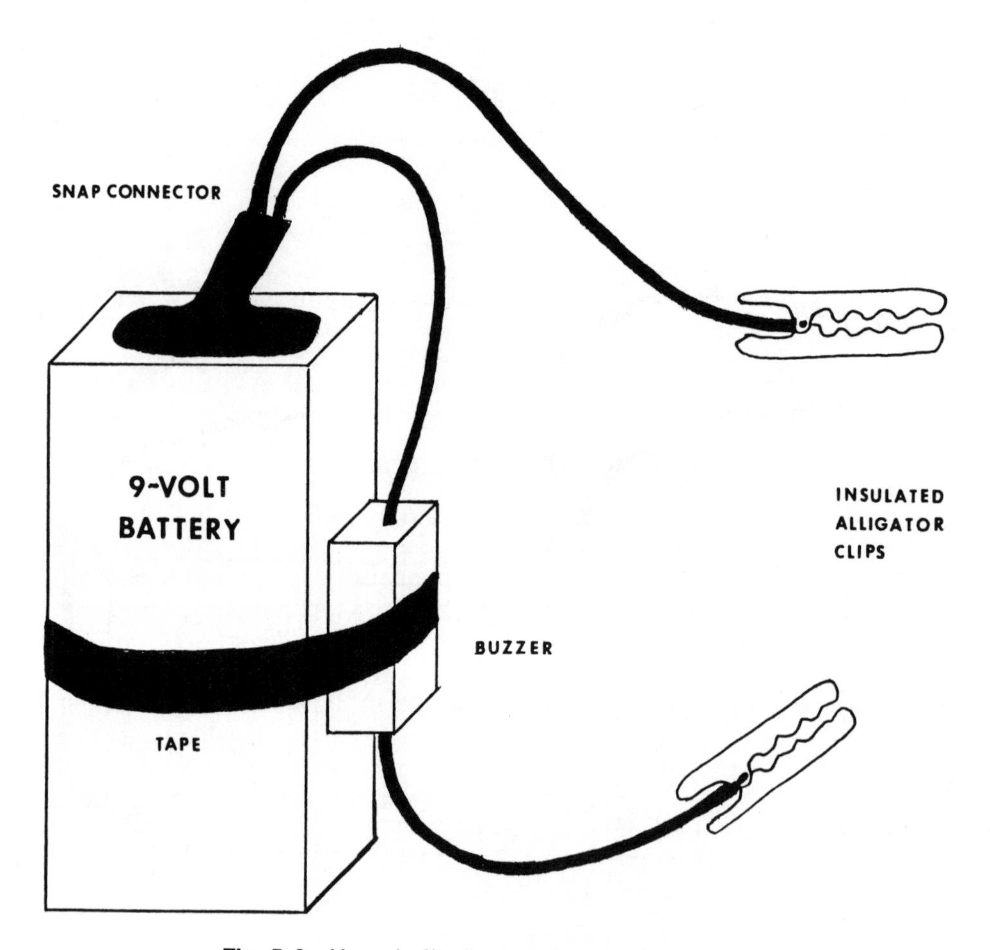

Fig. 5-8 Homebuilt short and ground tester.

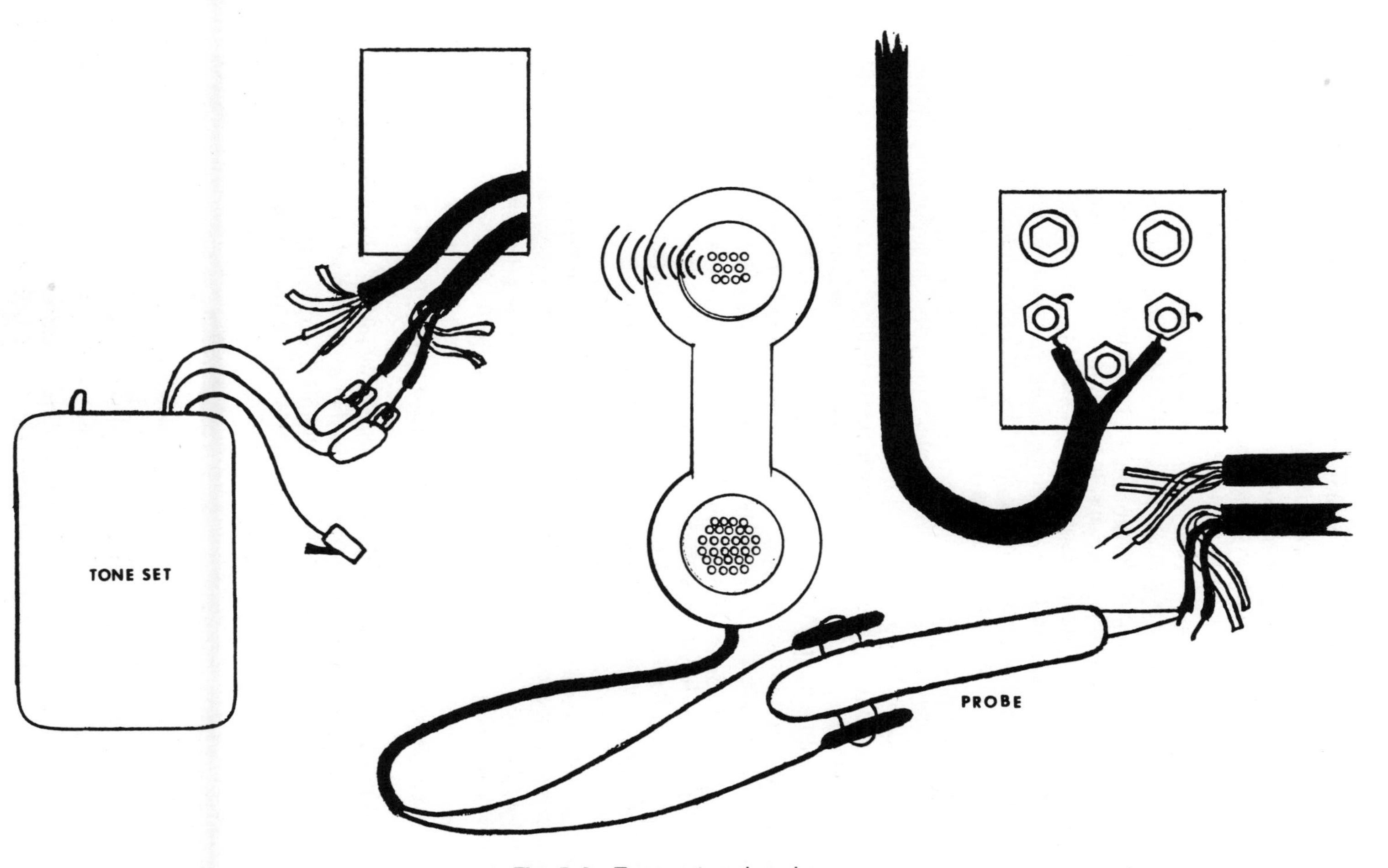

Fig. 5-9 Tone set and probe.

they make it possible to identify one pair from among many. With the tone set, a tone is put on the telephone cable pair at one end. With the probe and test set, you will then be able to find the tone on the other end. To send tone over a pair, simply connect one lead to the tip wire and the other lead to the ring wire and place the selector switch on tone. To hear the tone, connect the test set leads to the probe and hold the pointed end next to the pairs. When a loud tone is heard, you will have identified the desired pair.

The tone set also doubles as a short tester if you connect the tone set leads to the tip and ring wires. Set the selector switch to resistance and if the pair has a short, the tone set will indicate this with a fast beeping noise.

Most tone sets also have a polarity tester. To check for polarity, simply plug in the attached modularized cord to your phone modular connector. If a green light comes on, polarity is correct; if a red light comes on, polarity is reversed.

The tone set and probe are available at telephone company equipment suppliers, but if you decide to make your own, all you need are a transistor radio with a built-in earplug and two insulated alligator clips (Figure 5-10). Take the earplug and cut it off from the earplug extension cord. Attach the two insulated alligator clips to the ends of the cord. Select a radio station, plug in the earplug extension cord, and then attach the alligator clips to the pair: one alligator clip to the ring wire and one to the tip. When the homebuilt tone set is connected, the radio station broadcast will be sent over the pair. At the other end, you will need to use your test set as a probe by connecting its leads to all pairs until you hear the radio station.

An alternative to the homebuilt tone set is to use your volt and ohms meter with the selector switch set to ohms. Have someone else put a short on the pair and take it off about every two seconds. This way, you can take the meter to the location where you want to identify the pair and check all the pairs as you would for shorts. When you come across a pair with an intermittent short, you know it is the one you have been looking for.

Isolating Procedures for Telephone Circuit Trouble

In order to understand the isolation of telephone circuit trouble, you need to be able to identify both sides of a telephone circuit—the central office side and the telephone instrument side. If you break open the telephone circuit at any point from the central office to the telephone instrument, the side traveling back toward the central office will be the central office side, and the side traveling toward the telephone instrument will be the telephone instrument side. Figure 5-11 illustrates a telephone circuit broken open at three different locations. The first shows a telephone circuit broken open at the house protector; the second at a phone modular connector; and the third at the line cord.

Before you proceed to determine at which side of the demarcation point the trouble is, disconnect all telephone equipment from your telephone circuit. If the telephone circuit trouble is due to telephone equipment problems, the trouble will clear when the equipment is disconnected. Then reconnect the telephone equipment one piece at a time, checking

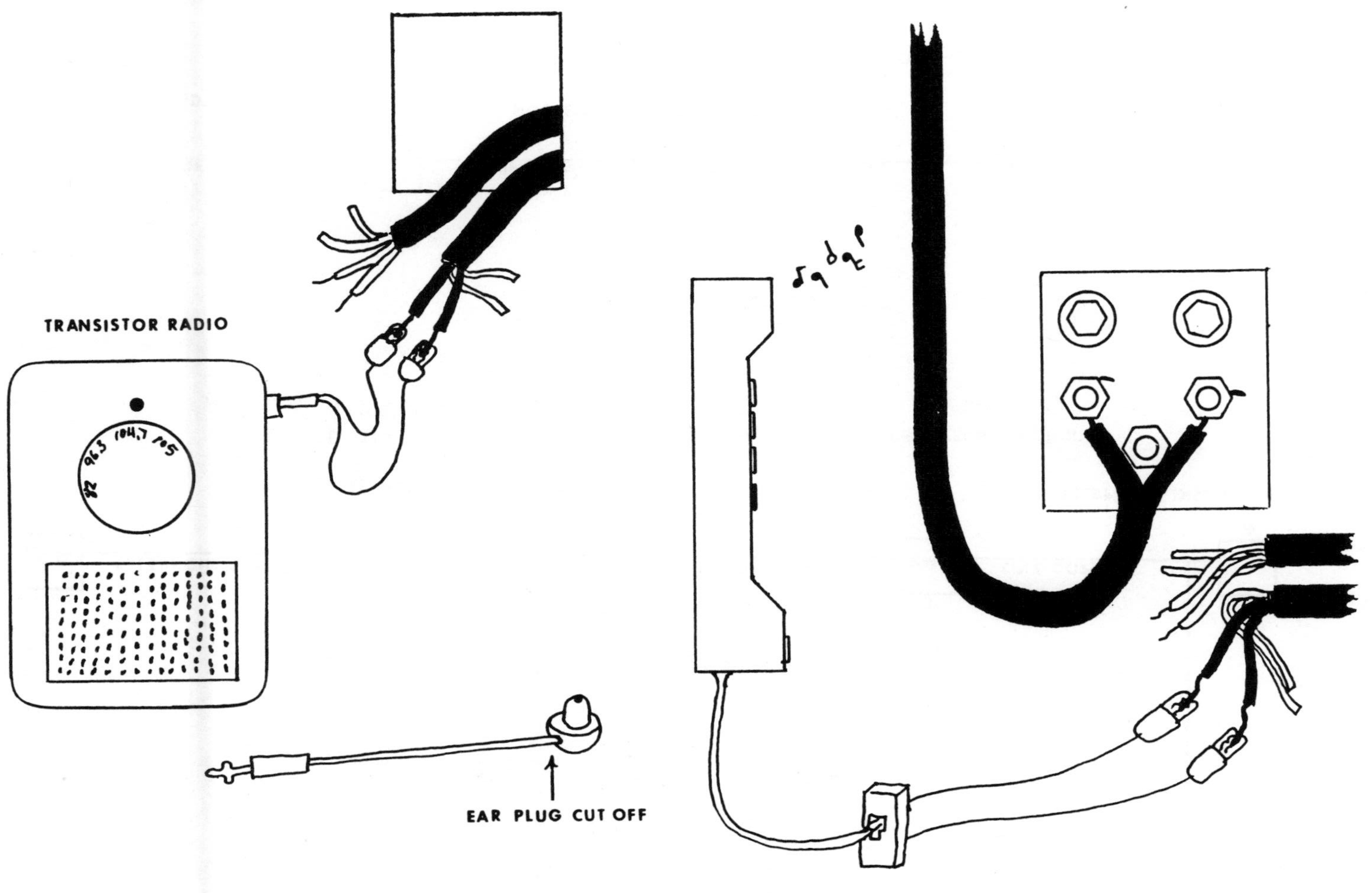

Fig. 5-10 Homebuilt tone set.

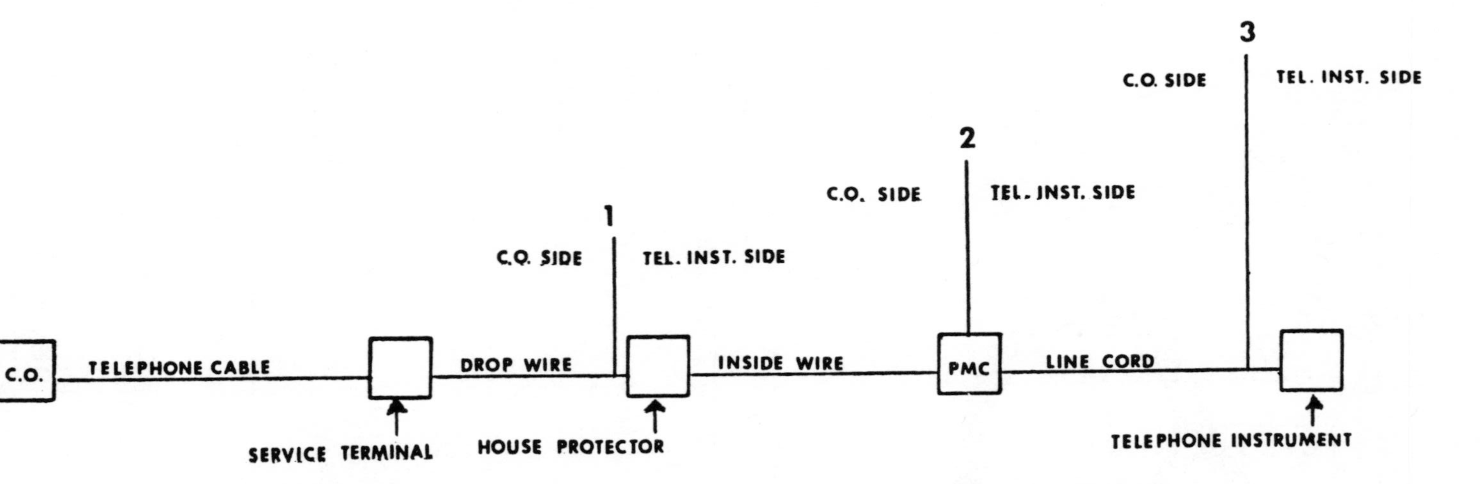

Fig. 5-11 The two sides of the telephone circuit—central office versus telephone instrument.

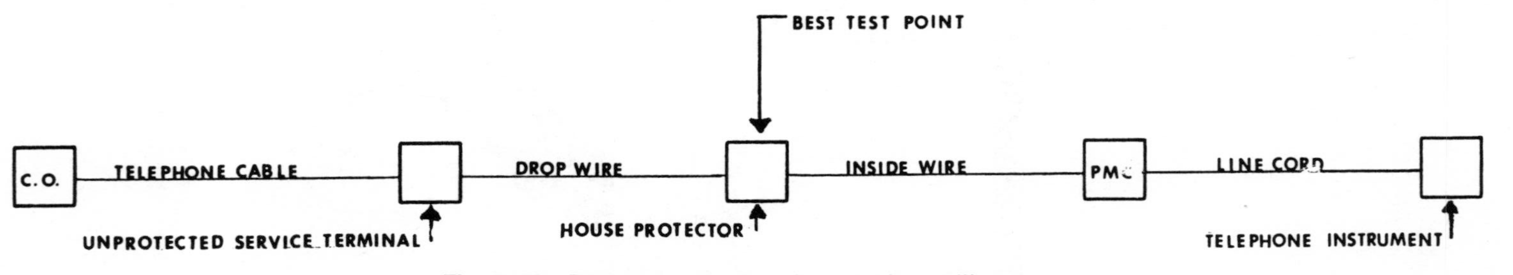

Fig. 5-12 Best test point for single-unit dwellings.

the telephone circuit after each piece is reconnected. This method will isolate the trouble to a specific piece of telephone equipment.

If the trouble is still on the telephone circuit after you perform this test, it will be one of two things: the trouble is not with the telephone equipment and is somewhere else in the telephone circuit, or the trouble is with the telephone equipment, but your telephone circuit has been disconnected. (Remember that automatic testing procedures may have disconnected your telephone circuit from the central office. So even when the telephone trouble is repaired by your efforts, the dial tone and line voltage may not reappear on the telephone circuit. In this situation, it will be essential to use test equipment to prove that telephone circuit trouble does exist and to prove that you have repaired it.)

The next step is to isolate the trouble to either the central office side or the telephone instrument side of the demarcation point (Figure 5-1). However, as explained earlier, you may not be able to test at your demarcation point, so you will want to choose a testing point as close as possible to the demarcation point.

The Test Point

When isolating telephone circuit trouble for single-unit dwellings, the test point will be the house protector. Depending upon your local telephone company's policy, it may also be the demarcation point (see Figure 5-1 and Figure 5-12). For multiunit dwellings, the test point may be at a protected service terminal with a telephone company–placed demarcation block for customer testing (Figure 5-13), at a 42A connector block demarcation and testing point (see Figure 4-5), or at the number one PMC (see Figure 4-4).

If your test point is located next to the protected service terminal, you will more than likely need to tone your pair to the protected service terminal in order to be able to find it. Although inside wire pairs will sometimes be marked to identify the unit from which it comes, it would be wise to verify the pair anyway since the markings are occasionally incorrect.

Once you have located the test point and checked your telephone equipment, break open the telephone circuit at the test point and check the central office side of the circuit for the five telephone circuit troubles. If you isolate the trouble to that side of the test point—and the test point is also your demarcation point—the local telephone company will be responsible for repairing the circuit at no additional expense to you. If, instead, you isolate the trouble to the telephone instrument side of the test point, you will need to do some more testing to specifically isolate the trouble.

By testing each inside wire pair (if you have the home-run hookup; see Figure 3-13 in Chapter Three), you will be able to isolate the trouble to a specific inside wire. If you have only one inside wire pair in an in-series hookup (see Figure 3-12 in Chapter Three), you know the trouble is isolated to that particular wire.

If the trouble is with the inside wire, take the extra pair within the inside wire and exchange it with the one in trouble. You will then need to go to all the phone modular connectors and make the exchange there, too.

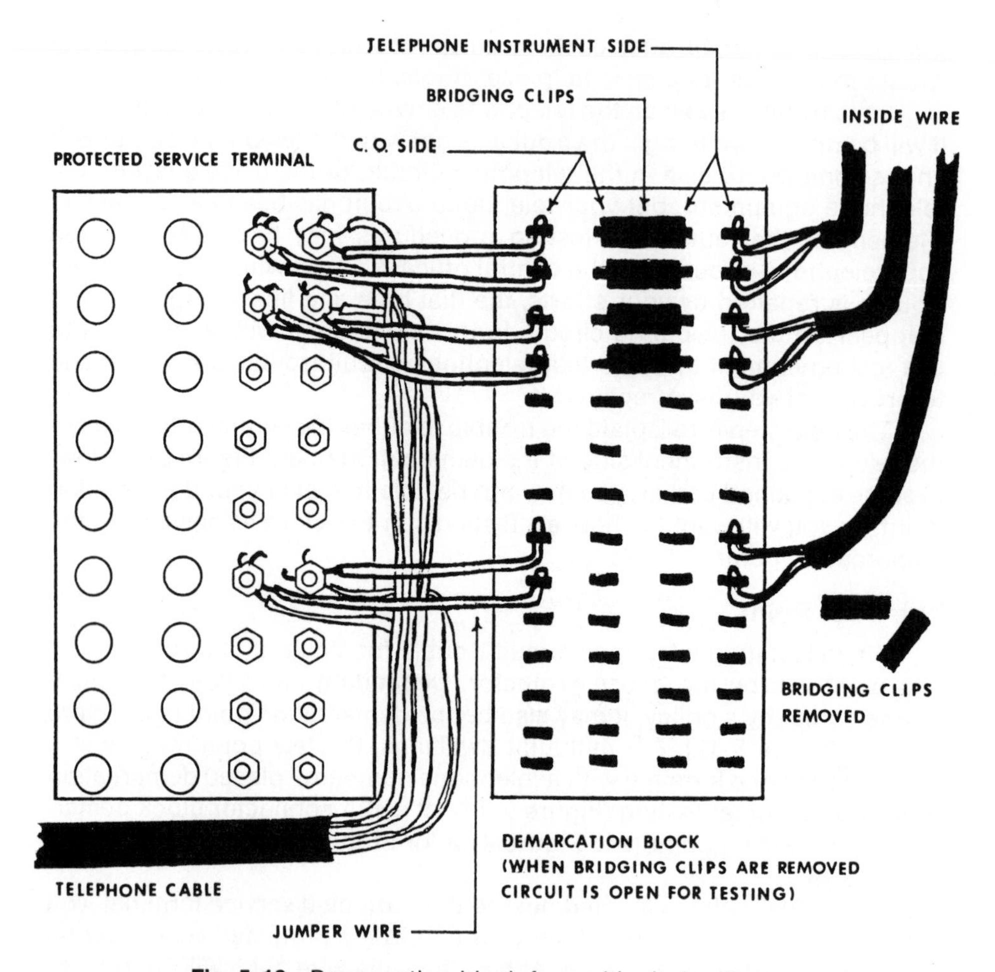

Fig. 5-13 Demarcation block for multiunit dwellings.

If the inside wire is clean, it means you have an open telephone circuit. Again, take the extra pair within the inside wire and exchange it with the one in trouble. You will then need to go to all the phone modular connectors and make the exchange there, too.

As you make the rounds exchanging the inside wire pair, test all phone modular connectors for a poor connection. Inspect the pins inside the phone modular connector plug for corrosion or crossed pins; check for broken or shorted lead wires that connect the pins to the terminal lugs; and check the inside wire pair where it is connected to the terminal lugs. After disconnecting the bad pair and before connecting the extra pair, test the phone modular connector with the ohmmeter. Connect one test lead to the 3-red terminal lug and connect the other test lead to the 4-green terminal lug. If no short is indicated, take a flathead screwdriver and carefully—so as not to damage the pins—insert it inside the plug, shorting the wires to verify that the phone modular connector does not have an open circuit. If the test indicates that the trouble is with the phone modular connector, you will not have to exchange the inside wire pairs.

Troubleshooting Guide

This guide presents the twelve most common types of telephone troubles (relating to the instrument as well as the circuit), the reasons for each, and procedures for isolating and repairing the problem in the least amount of time possible.

Before you can effectively use this troubleshooting guide, it is essential that you have a good understanding of "Five Basic Types of Telephone Circuit Trouble" and "Isolating Procedures for Telephone Circuit Trouble" in this chapter.

Before you worry about specifics, you can determine whether or not the problem is with the telephone instrument. To do this, plug the phone into a phone modular connector that you are sure works (e.g., a neighbor's). If the phone works, the problem is in the station wiring or local network; if it doesn't work, the problem is with the telephone instrument. Another method is to plug a phone that you are sure works into the phone modular connector in question; if the instrument works, the problem is with the other phone.

Next, find your problem in the following list and follow the directions given.

1. Unable to receive calls

You have a dial tone and can place outgoing calls but are receiving calls that are not for you.

Causes

Your telephone circuit is connected to the wrong telephone number. This type of telephone trouble typically occurs on the initial installation of your telephone service, because the installation technician made an error in the field and connected you to another telephone number. Thus it is a good idea to make sure you have the right number before the technician leaves. In addition, your telephone circuit may become connected to another telephone number even after you have had service for quite some time. This can happen when telephone company cable splicers make a splicing error that connects your telephone circuit to another circuit harboring a different telephone number.

What to do

Because the trouble is in the field—somewhere between your protected or unprotected service terminal and the central office—you can simply call your local telephone company repair office and tell them the telephone number connected to your phone is not your telephone number. You can shorten the time the telephone company will need to isolate the trouble by providing them with the number that is incorrectly connected to your phone. (The next time the telephone rings, ask the caller what number he or she dialed. To check, call this number from your home telephone; if this is the number your phone is connected to, you will hear a busy signal or a low tone beeping—a ring-back indicator [when you hang up, your phone will ring]. Note that not all areas have the same method for ring back; some areas have a three- or four-digit code not related to the telephone number dialed. Another way to find the telephone number connected to your phone is to call the operator. Many operators have a computer terminal screen that displays the telephone number of the incoming caller.)

2. Unable to receive calls

You have a dial tone and can place outgoing calls but are not receiving any incoming calls.

Causes

The ringer inside your telephone instrument is not working.

What to do

Open the telephone instrument and see if the wires from the ringer to the network are loose. To make certain that your ringer is bad, have a friend call you at a specified time and let the phone ring for about a minute. If the ringer does not sound and your friend is on the other end, you have a bad ringer.

3. Unable to receive calls

You have a dial tone and can place outgoing calls, but the ring is too short and when you answer the telephone no one is there. This situation is often accompanied by static.

Causes

Somewhere in your telephone circuit is a slight short, typically caused by wet conditions. A slight short will become a solid short when the incoming call occurs because the voltage from the ringing generator is higher than the voltage on the telephone line when the phone is on the hook or when you are talking on it. The solid short causes the incoming call to be disconnected.

What to do

Since the trouble may be anywhere in your telephone circuit, first isolate the trouble at the demarcation point (see "Isolating Procedures for Telephone Circuit Trouble" in this chapter). If the trouble is isolated to the central office side of the demarcation point, call your local telephone repair. If the trouble is isolated to the equipment side of the demarcation point, look for a phone modular connector or modular line cord that has a blue powdery substance stuck to it; this will indicate the inside wire or phone modular connector with the slight short. The blue powdery substance is caused by electrolysis when wet conditions and slight shorts exist. Inspect inside wire runs for nicks that may be exposed to wet conditions. Phone modular connectors with the blue powdery substance should be replaced; modular line cords with the substance can be cleaned with a stiff brush and alcohol.

4. Intermittent dial tone

Causes

The trouble can be anywhere in your telephone circuit or instrument. Somewhere in the telephone circuit may be a loose connection that occasionally breaks contact.

What to do

You first need to isolate the trouble at the demarcation point (see "Open Telephone Circuit" and "Isolating Procedures for Telephone Circuit Trouble" in this chapter). If you isolate the trouble to the equipment side of the demarcation point, check your telephone instrument, and then look for loose connections at phone modular connectors and at the house protector or protected service terminal. If nothing is found, inspect inside wire runs for kinks or pinched spots.

5. Intermittent static

Causes

The problem can be anywhere in the telephone circuit; there may be bad telephone cable pairs, loose connections at phone modular connectors, or loose connections in the telephone instrument.

What to do

Intermittent static is the most difficult telephone trouble to isolate and repair because the cause can be anywhere in your local telephone network or other telephone networks you call. The first step is to determine the origin of the static. Are you getting static on incoming, outgoing, local, *and* long distance calls? If so, the trouble is with your local telephone network and you will need to isolate at the demarcation point. Check contact points inside the telephone instrument, check modular line cord connections, and check the house protector or protected service terminal for spider webs and loose connections. Are you getting static on long distance calls only? The trouble will be at the central office and beyond—call the local telephone company repair. Are you getting static on local incoming calls only? The trouble is with the central office where the call originated from—tell the caller to contact his or her local telephone company repair. Static on long distance calls only may be caused by poor trunking telephone cables and connections and is difficult to report and have repaired. The long distance company that you are using will have maintenance crews that monitor trunk lines continuously and make repairs when the trouble is located. (The promptness of repair is what makes the difference in the quality of the long distance company you use.)

6. No dial tone

Causes

This problem can be caused by a bad telephone instrument or a bad telephone cable pair.

What to do

Check your telephone instrument. If the trouble is not with the instrument, you will need to isolate at the demarcation point (see "Isolating Procedures for Telephone Circuit Trouble" in this chapter).

7. Unable to be heard

On both incoming and outgoing calls the party on the other end cannot hear you.

Causes

The trouble is with the transmitter unit in the handset of your telephone instrument.

What to do

You will need to check the connections to the transmitter unit; if the connections are good, the transmitter unit itself will need replacing.

8. Unable to hear

On incoming and outgoing calls the party on the other end can not be heard by you.

Causes

The trouble is with the receiver unit in the handset of your telephone instrument.

What to do

You will need to check the connections to the receiver unit; if the connections are good, then the receiver unit itself will need replacing.

9. Unable to call out

You have a dial tone and can receive incoming calls.

Causes

When you are unable to call out, the trouble is with the telephone instrument.

What to do

For rotary dialing telephone instruments, the trouble is with the rotary dial connections or the hook switch connections. For touch tone telephone instruments, the trouble is with either the touch pad connections or hook switch connections or is a reverse polarity with the telephone circuit.

10. Ringer doesn't ring loudly

Causes

A low ringer volume is usually caused by one of four things; the ringer volume adjustment is turned down, the tension spring on the ringer is bad, you have exceeded your REN of 5, or the local central office has changed its switching equipment with the result that your designated "ringer cycle" was changed.

What to do

First check to see if your volume adjustment is set on low. Next open the telephone instrument and inspect the ringer and ringer connections for loose or broken connections. Then add up the total RENs on all your telephone instruments. If these procedures do not reveal the trouble, call your local telephone company repair and see if your telephone was wired for a certain type of ringer cycle frequency. If it was and your local telephone company changed its switching equipment, you'll need to have your ringer wired for the new switching equipment.

11. Others on line

You pick up your telephone instrument and hear another conversation on the line. You check your local telephone company bill and there are charges to telephone numbers that you have not called.

Causes

This problem typically occurs when your telephone service is initially installed, because the cable pair that was assigned for your service is still connected to the station wiring of the home where the cable pair was previously being used. In addition, you may hear other parties on your line if your cable pair is being tapped from other service terminals also connected to your pair.

What to do

When you pick up your phone and hear another conversation, try to obtain the address of the people using your line. The address will help telephone repair people locate the exact spot where your cable pair is connected to some one else's station wiring. The telephone company assignment department will have records of where your cable pair is connected to other service terminals and will conduct a search at your request to see if your cable pair is connected to someone else's station wiring at one of these service terminals. In this situation, you will need to pay close attention to your telephone bill for telephone numbers you know you haven't called. Report these to the telephone company billing office. If the problem is legitimate, the telephone company will remove the charges.

12. Reverts to dial tone after calling

Calls you dial are not completed; the telephone line reverts back to dial tone.

Causes

There are two reasons for this type of trouble: the telephone instrument and the switching equipment at the local central office.

What to do

If you have proved that the trouble is not with the telephone instrument, then the trouble is with the central office. Call your local telephone company repair department.

Special Types of Telephone Trouble and Repair

If you happen to be connected to a special telephone circuit other than the one that connects your telephone instrument to the local central office, you need to be aware of special conditions that may apply if the circuit falls into trouble.

Field Exchange Circuits

If your telephone circuit is in a field exchange circuit (see Figures 2-2 and 2-3 in Chapter Two) and you have isolated the trouble to the central office side of the test point, telephone repair procedures may involve two or more central offices, which may be in two or more service areas. Unless you know specifically where your service originates, you may have difficulty locating and informing the proper personnel.

Repairing field exchange circuits that travel through two or more telephone company service areas may also be a complex issue unless you know who to contact and what specific information to give them. When having a field exchange circuit installed, be sure to find out how to contact repair service for the area in which the service originates and what central offices and service areas the circuit will travel through. Telephone company personnel may be just as confused as you if they do not realize that your telephone circuit is in a field exchange circuit.

Off-Premise Extensions

With the off-premise extension, you have two separate areas of responsibility. If you isolate trouble to the central office side of the test point at your primary location, it may still be possible that the trouble is caused from the telephone instrument side of the test point at the secondary location (see Figure 2-4 in Chapter Two). In order to accurately isolate circuit trouble to the central office side of the test point, you will also need to test at the secondary location.

Achieving Telephone Communications—Special Situations

At the House Protector

If you live in a single-unit dwelling and have isolated trouble to the telephone instrument side of the house protector—but it is not possible for you to repair the trouble at the moment—you can connect an emergency telephone as long as you still have a dial tone at the central office side. To do this, you will need a phone modular connector and enough additional inside wire to reach from the house protector to your home. Disconnect existing inside wire from the house protector and attach the additional inside wire. Run it inside your home and connect the phone modular connector, which creates an emergency telephone circuit (Figure 5-14).

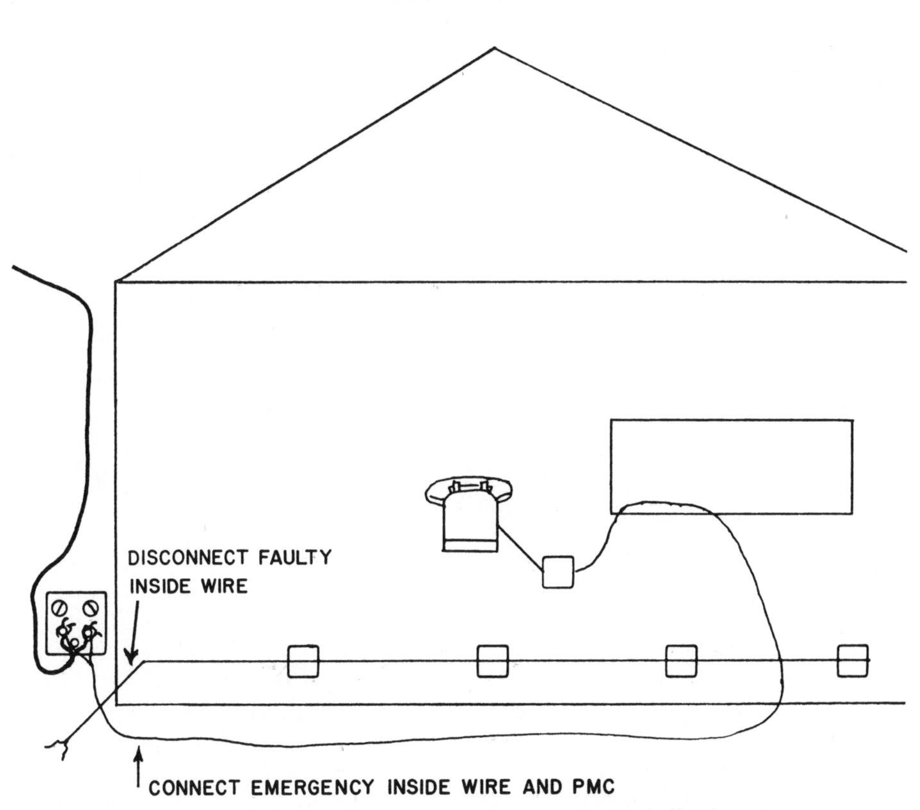

Fig. 5-14 Connecting an emergency circuit at the house protector.

Call Forwarding

If you live in a multiunit dwelling and the house protector emergency telephone circuit is not possible or you have isolated the circuit trouble to the central office side of the test point, call forwarding may be the best emergency solution. Ask repair personnel to contact the central office and have the switching personnel call-forward your telephone number to another number that would be agreeable to you. Use that number until the telephone circuit trouble is repaired.

Mobile and Cellular Mobile Telephones

If you live in a service area that provides poor telephone communication, you may consider mobile or cellular mobile telephone systems as an emergency backup. Since these two systems are not connected to your local central office by the telephone cable pairs the way your home telephone system is, you will not experience the same type of service problems.

Ham/Amateur Radio

Ham (or amateur) radio is another alternative for telephone communications in an emergency situation. Federal regulations governing amateur radio, which are enforced by the Federal Communications Commission (FCC), state that amateur radio cannot be used to make a profit but can be used for personal pleasure and for life-threatening emergencies.

With today's state-of-the-art amateur radio and radiotelephone communications, it is possible to place and receive calls on a small, hand-held radiotelephone with direct dialing from just about anywhere in the country. When comparing costs between mobile or cellular mobile telephones for strictly emergency needs, the amateur radio radiotelephone is about one-third the cost for the equipment and about one-tenth the cost to operate.

To become a licensed amateur radio operator, check with your local FCC Field Engineering Bureau Office or the American Radio Relay League to learn the requirements. These two sources will also be able to direct you to information sources that will supply exact details for amateur radio and radiotelephone communication within your area.

Marine Radio

If you live close to the ocean and within approximately 25 miles from a harbor, you may want to consider the marine radiotelephone as an alternative for emergency telephone communications. Although the marine radio is designed for ship-to-shore and ship-to-ship communications, within the marine radio frequencies there are a few channels designated for marine radiotelephone use. Ordinarily, the marine radio operator would require that you identify yourself with a station license number and the name of the ship from which you are calling. However, in an emergency situation when life is in danger or substantial damage to property is imminent, radio regulations and procedures are dropped in order to provide immediate help.

Reporting Trouble to the Local Telephone Company

When reporting trouble to the telephone company, refer to the telephone installation checklist in Chapter Six because much of the same information is applicable in a repair situation.

Briefly, ensure that telephone company repair technicians have proper access to all points in your telephone circuit. If you live in a multiunit dwelling and suspect that inside wiring needs replacing, have the Building Owner's Consent Form ready. If you have special telephone circuits, such as off-premise extensions, field exchange circuits, or alarm circuits, notify the telephone company repair department about them. If your telephone circuit is continuously in trouble, keep a telephone trouble log. Each time you report trouble, record the following information:

1. Date and time of trouble.
2. Date and time of reporting trouble.
3. Name of individual at the local telephone company to whom you reported the trouble.
4. Type of trouble reported.

5. Name of telephone repair technician dispatched to repair the trouble.
6. Type of trouble found by the repair technician.
7. Date and time that trouble was repaired.

When calling for repair work, ask to speak to the service repair department supervisor and then have that supervisor contact the supervisor for the repair technician department. If after contacting the telephone company's management, your circuit trouble is still not repaired, contact the Public Utilities Commission (PUC) to put pressure on the telephone company. Since the PUC regulates the service of the local telephone company, it will require that you have your facts correct concerning your continuous telephone circuit trouble or else it will give the benefit of the doubt to the company. Your telephone trouble log will give the PUC the justification necessary to apply pressure on the local telephone company.

CHAPTER SIX

Getting Along with the Local Telephone Company

Home Telephone Installation Checklist

It is a good idea to conduct a survey of your home telephone system before contacting the local telephone company to order a telephone service. Use the following checklist as a general guide.

☐ **SINGLE-UNIT DWELLINGS**

1. The Unprotected Service Terminal
    a. Will telephone installation people have proper access?
    b. Are there poor conditions you need to report?
2. The Drop Wire
    a. Is there adequate drop wire for your needs?
    b. Is existing drop wire in poor condition and in need of replacement?
    c. Will existing drop wire require rerouting?
    d. If it is buried drop wire, will you need to have it located?
3. The House Protector
    a. Does your house have a house protector?
    b. Will an existing house protector require relocating?
    c. Do you want a multistation house protector as opposed to single station house protectors for two or more telephone numbers?
    d. Do you have an old type of nonfused house protector that needs replacing?
    e. Are existing house protectors in good condition?

☐ **MULTIUNIT DWELLINGS**

1. The Protected Service Terminal
    a. Will telephone installation people have proper access?
    b. Are there poor conditions you need to report?

☐ **SINGLE-UNIT AND MULTIUNIT DWELLINGS**

1. Inside Wire and Phone Modular Connectors
   a. Are there any hard-wired or four-prong phone jacks that need converting to phone modular connectors?
   b. Are existing phone modular connectors adequate; are their locations suitable for your needs?
   c. Is existing inside wire sufficient for installing extension phone modular connectors or additional telephone numbers if required?
   d. Is existing inside wire in poor condition and in need of replacement?
   e. Do you need a wall phone modular connector installed?
   f. Is a Building Onwer's Consent Form required to install or replace inside wire?
   g. Will you want phone modular connectors wired to accept two different telephone numbers?
   h. If your home has a crawl space and installation of inside wire utilizes the run-under method, is the crawl space clean, dry, and accessible?
   i. Are there any unusual surfaces that inside wire or phone modular connectors must be attached to?
   j. Are there any unusual circumstances an installation person would have to deal with?
2. Telephone Services and Instruments
   a. Do you have the right telephone instrument for either touch tone or rotary dialing?
   b. Do your telephone instrument's ringer equivalent numbers (REN) total more than five?
   c. Do you know the FCC registration numbers for your privately owned telephone instruments?
   d. Will your telephone instruments require line cord changes for phone modular connectors wired to accept two different telephone numbers?
   e. If your answering machine is Calling Party Controlled, does the local telephone company's central office provide the proper signal to operate it?
   f. Will you require an RJ-31X phone equipment jack installed for alarm, data, or long-distance discount dialing services?
   g. Have you gone through this checklist for off-premise extensions locations?
   h. Do you have telephone instruments with lighted dials or touch pads that require an external power source?

Types of Telephone Installations

To place an order for the installation of telephone service, contact your local telephone company customer service office. The following sections present the major types of service installations.

The Home-Run

The home-run installation, the least expensive and the least time-consuming, is when the telephone circuit between your home phone modular connectors and the local telephone company central office already exists. When a home-run installation is possible, the central office receives your order for service and from it will know what telephone cable pair to connect to its switching equipment.

The set of circumstances that typically make it possible for telephone service to be installed with a home-run connection are: when you move into a house soon after previous occupants have moved out but before the assignment department has reassigned the cable pair for another service; when all home station wiring is serviceable and acceptable for your needs; and when the existing telephone circuit within the local network is in good condition.

Field Visit

The field visit is a slightly more expensive installation and is required when the telephone circuit between the central office and your house is not complete or needs changing because the cable pairs have gone bad. For a field visit, the installation technician will be required to connect your telephone circuit cable pair at one or two of the following locations: a cross-connect terminal, protected service terminal, or an unprotected service terminal.

Premise Visit

The premise visit is required when the telephone circuit cable pair needs to be connected to your home telephone station wiring. For single-unit dwellings, this usually means installing a house protector and connecting or installing a drop wire then connecting this to existing inside wire with phone modular connectors. With multiunit dwellings that require a field visit, a premise visit is usually required, too, because protected service terminals are seldom labeled with the address or apartment number corresponding to the inside wire. The telephone technician will need to access your house in order to tone your inside wire back to the protected service terminal for proper identification.

Home Installation Visit

The home installation visit, required when you wish the local telephone company to install inside wire and phone modular connectors for you, is the one expense you can control yourself since you can either install your own inside wire and phone modular connectors or elect to have a private company do it. The most expensive installation for normal telephone service includes a field visit, premise visit, and charges for the number of phone modular connectors installed by the telephone technician during a home installation visit.

Special Telephone Circuits

If your home telephone system requires the connection of special telephone circuits, such as field exchange circuits, off-premise extensions, data circuits, or ring-down circuits, there will be additional installation expenses and sometimes additional time required to install them.

Many times your order for a special telephone circuit will be routed to the engineering department, which will design the circuit to determine if it can be constructed and to ensure that it works properly. If the special circuit crosses from one service area network into another, that engineering department must also be involved in the installation process.

The cost and time associated with the installation of special telephone circuits depend on the type of service and the distance involved in connecting the circuits. Since it is possible to have various installation technicians from many local networks and service areas involved with the connection, the installation could become time-consuming and costly.

Installation Rescheduling

The telephone field technician receives on the average eight to ten installation orders to complete in a day. The experienced technician reviews each service order, checking for completeness and special instructions.

Many times these appointments require rescheduling because of incomplete or missing special instructions. Incorrect or incomplete addresses also result in cancellations. If you live at an address that is difficult to locate, leave special instructions along with a telephone number where you (or your representative) can be reached.

If you have a friend or neighbor meet the technician, make sure that person knows exactly what you want or the technician may reschedule his or her visit. Also, clear all visits through the management office, gate house, or other security area if you live in a multiunit dwelling.

In addition, dogs are a major reason why field technicians reschedule installation or repair appointments. Injury—and law suits—may result if a technician is bitten, so many will refuse to enter the premises if a dog is running loose.

Index

Page numbers in **boldface** refer to illustrations.